FRONTIERS IN QUANTUM DOTS RESEARCH

FRONTIERS IN QUANTUM DOTS RESEARCH

MARCO L. VORONIN
EDITOR

Nova Science Publishers, Inc.
New York

For permission to use material from this book please contact us:
Telephone 631-231-7269; Fax 631-231-8175
Web Site: http://www.novapublishers.com

Library of Congress Cataloging-in-Publication Data

Frontiers in quantum dots research / [edited by] Marco L. Voronin.
p. cm.
Includes bibliographical references and index.
ISBN 13: 978-1-60021-057-0
ISBN 10: 1-60021-057-0
1. Quantum dots--Research. I. Voronin, Marco L.
QC611.6.Q35F76 2006
621.3815'2--dc22 2006006138

Published by Nova Science Publishers, Inc. ✣ New York

CONTENTS

PREFACE

A quantum dot is a particle of matter so small that the addition or removal of an electron changes its properties in some useful way. All atoms are quantum dots, but multi-molecular combinations can have this characteristic. In biochemistry, quantum dots are called redox groups. In nanotechnology, they are called quantum bits or qubits. Quantum dots typically have dimensions measured in nanometers, where one nanometer is 10-9 meter or a millionth of a millimeter. The fields of biology, chemistry, computer science, and electronics are all of interest to researchers in nanotechnology. Other applications of quantum dots include nanomachines, neural networks, and high-density memory or storage media. Research is being carried out on nano-crystals, self-assembled dots, and gated structures.This book presents leading-edge research from around the world.

Chapter 1 addresses issues associated with physics and technology of long-wavelength diode lasers based on self-organized quantum dots (QDs). The authors review recent results on epitaxial growth, studying the structural and optical properties of QDs, fabrication and characterization of edge- and surface-emitting diode lasers with special attention paid to the relationship between structural and electronic properties of QDs and laser characteristics. Performance of the state-of-the-art QD lasers is described. We mostly focus on pseudomorphic quantum dots suitable for 1.3μm-emitting lasers grown on GaAs substrates. In addition, very recent results on 1.5μm QD lasers based on metamorphic approach are also discussed. The authors wittingly restrict the book content to InGaAs material system as the most developed and practically exploited.

In the last two decades, the field concerning quantum dots, and, in general, low dimensional semiconductor structures has experienced great advances from both experimental and theoretical points of view. At present, these structures constitute one of the most suitable scenarios to investigate quantum phenomena in condensed matter. As described in chapter 2, zero-dimensional structures (quantum dots) provide an ideal system to achieve this purpose due to the intrinsic quantum nature of the dynamics characterizing the particles confined; for instance, shell structure of the quantized levels constitutes an illustrative example and the origin of the terminology 'artificial atoms' because of the similarity with natural atoms.

As explained in Chapter 3, the charge density and pair correlation function of three interacting electrons confined in a two-dimensional circular quantum dot are calculated by the exact diagonalization of the Hamiltonian. The circular quantum dot is characterized by the hard wall type confinement potential to obtain the observable effects of the electron-electron interactions on the electron distribution within the quantum dot. For the circular quantum dot geometry, in addition to the total spin S, S_z and the z- component of the total orbital angular momentum M are good quantum numbers. Furthermore, both single particle energy level and matrix element of the Coulomb potential are independent of the spin and sign of m. All of these symmetry properties cause us to interest in many electron states having fixed values of M and S_z for a given electron number, N. Also the authors restrict ourselves to those states for which $S_z = S$ to work out the relevant quantities in detail. In the analytical and numerical calculations of the charge density and pair correlation function, they assume spin polarized three electrons with vanishing total orbital angular momentum, $M = 0$. Variation of the charge density with radial distance as well as that of the pair correlation function with azimuthal angle together with their size dependence show clearly the atomic limit and the formation of the Wigner molecule. The Wigner-molecular arrangement of three electrons occurs in the form of an equilateral triangular structure as the size of the dot is increased. In this quasi-classical configuration, the electrons minimize their electrostatic energy only by arranging themselves at the vertices of an equilateral triangle.

Quantum dot (QD) structures have attracted considerable attention in the past decades because they not only address an important issue in basic physics, but also find potential and practical applications in many fields. II-VI materials have long been studied for their larger band gap, higher exciton binding energy and larger polarity than their III-V counterparts. If combining the advantages of QD structures and those of II-VI materials, a charming prospect can be expected. Under this circumstance, many groups began to study the II-VI QDs. Some noteworthy progresses have been obtained after several years' research. In Chapter 4, the authors will summarize these progresses, and the summarization will be divided into five parts: 1. The preparation of II-VI QDs. In this part, they will demonstrate the two epitaxial growth modes of QDs, Stranski-Krastanow (S-K) mode and Volmer-Weber (V-W) mode. 2. The formation mechanism of S-K mode QDs. The authors depict the formation mechanism of QDs grown in S-K mode, in which the energy reduction induced by atom diffusion will be considered. 3. The optical properties of II-VI QDs. 4. Progress in the practical applications of II-VI QDs. The light emitting devices and lasing diodes based on CdSe, CdS QDs will be summarized in this part. 5. The prospect of II-VI QDs. In this part, the unresolved problems in II-VI QDs are addressed, and their possible solutions are also dealt with.

In: Frontiers in Quantum Dots Research
Editor: Marco L. Voronin, pp. 1-49

ISBN: 1-60021-057-0

Chapter 1

LONG-WAVELENGTH QUANTUM-DOT LASERS

Alexey E. Zhukov** and Victor M. Ustinov
A.F. Ioffe Physico-Technical institute, Russian Academy of Sciences
26 Politekhnicheskaya, St. Petersburg, 194021, Russia
Alexey R. Kovsh and Nikolai N. Ledentsov
NSC-Nanosemiconductor GmbH
Joseph-von-Fraunhofer-Str.13, 44227 Dortmund, Germany

Abstract

This chapter addresses issues associated with physics and technology of long-wavelength diode lasers based on self-organized quantum dots (QDs). We review recent results on epitaxial growth, studying the structural and optical properties of QDs, fabrication and characterization of edge- and surface-emitting diode lasers with special attention paid to the relationship between structural and electronic properties of QDs and laser characteristics. Performance of the state-of-the-art QD lasers is described. We mostly focus on pseudomorphic quantum dots suitable for 1.3μm-emitting lasers grown on GaAs substrates. In addition, very recent results on 1.5μm QD lasers based on metamorphic approach are also discussed. We wittingly restrict the book content to InGaAs material system as the most developed and practically exploited.

Keywords: Quantum dots, diode laser, VCSEL, 1.3-1.55 μm wavelength

1. Introduction

Diode lasers are widely used as cost-effective and compact sources of coherent light in many applications including datacom/telecom, optical storage, pumping, printing, etc. Depending on the specific application, a diode laser should provide a certain wavelength of the emission. In particular, the 1.3-1.6 μm spectral band satisfies the demand of high-speed optical data

* E-mail address: Zhukov@beam.ioffe.ru; phone/fax: +7-812-247-9132/3178

transmission exploiting minimal dispersion (1310nm) or minimal attenuation (1550nm) of a standard silica fiber.

Till recently this spectral range has been out of the well developed technology of GaAs-based diode lasers due mostly to lack of suitable material of the laser active region. A GaAs substrate does not have any lattice-matched alloy that covers the required bandgap interval. At the same time GaAs-based structures have several advantages over InP-based counterparts. These include better electron confinement owing to larger bandedge discontinuities in these materials, better thermal conductivity, high refractive index contrast in Al(Ga)As/GaAs pairs. All this would make it possible a realization of uncooled lasers and surface-emitting lasers with advanced characteristics.

It is usually considered that a search for the GaAs-based system suitable for 1.3-μm application was pioneered by Kondow by introducing InGaAsN materials [1]. Almost simultaneously a potential of three-dimensional InGaAs islands, the so-called self-organized quantum dots (QDs), for long-wavelength applications on GaAs has been demonstrated [2]. It was found that redistribution of a two-dimensional film into three-dimensional islands leads to significantly longer wavelength emission achievable on a GaAs substrate.

Development of long-wavelength QD lasers is in tight connection with a long-term evolution of semiconductor lasers that exploit effects of size quantization. The advantage of use of the size quantized heterostructures in lasers comes from the increase of the density of states near the band edges. This original idea belongs to Dingle and Henry [3]. If that would be implied as an active region of a diode laser, this will result in a concentration of the non-equilibrium carriers in a narrow energy range near the bottom of the conductance band and the top of the valence band. This should increase effective gain and decrease temperature effects on device performance.

QD is the ultimate example of a size quantization which represents a semiconductor crystal with a size of several nm inserted into a larger bandgap semiconductor matrix. A quantum dot laser has been originally proposed to provide extremely low and temperature-independent threshold current owing to zero-dimensional density of states in the laser active region [4]. Since the pioneering work of Goldstein et al in 1985 [5] the foremost results have been achieved using spontaneous formation of three-dimensional islands in strained layer epitaxy.

To date reliable quantum-dot lasers with low threshold, high efficiency and high temperature stability operating in the 1.3-μm spectral range have been realized. Also, a possibility to extend the spectral range of GaAs-based emitters to 1.5-μm has been recently demonstrated. In this chapter we review recent results on epitaxial growth, studying the structural and optical properties of QDs, fabrication and characterization of edge- and surface-emitting diode lasers with special attention paid to the relationship between structural and electronic properties of QDs and laser characteristics. The chapter is organized as follows. Sec. 2 is devoted to the principles of quantum dot formation using self-organization phenomena highlighting the specific approaches which are applied to extend the long-wavelength limit of QD emission up to 1.3 μm. In Sec. 3 we mostly focus on 1.3μm-emitting QD lasers grown on GaAs substrates. Performance of the state-of-the-art QD lasers is described. Finally, in Sec. 4 very recent results on 1.5μm QD lasers based on metamorphic approach are discussed.

2. Self-organized Quantum Dots: Fabrication and Properties

A lot of efforts were undertaken to develop a fabrication technology of semiconductor quantum dots suitable for device applications. Since the mid of 1990s there has been considerable work on direct synthesis of semiconductor nanostructures by applying the phenomenon of island formation in lattice-mismatched systems. Semiconductor QDs exhibiting properties sufficient for laser applications are currently fabricated using the effect of spontaneous transformation of the growth surface at initial stages of strained-layer heteroepitaxy, a process called the Stranski-Krastanow growth mode.

Epitaxial deposition of a semiconductor material on top of another one that has a lattice constant several percent smaller, e.g. InAs on GaAs, leads to appearance of elastic strain in the structure. First, such a growth proceeds in two-dimensional way, forming a planar "wetting layer". However, as soon as the layer thickness exceeds some critical value then the strain relaxation takes place and formation of well-ordered three-dimensional array of nanoscale islands (so called quantum dot array) occurs on the top of the thin wetting layer. When those islands are covered with subsequent GaAs overlayer then one has an array of InAs insertions inside of the GaAs matrix. This mechanism of QDs formation in Stranski-Krastanow mode is illustrated in Figure 1.

Overall, Stranski-Krastanow growth makes it possible to prepare extremely small quantum dots in a maskless process without lithography and etching. This capability makes it a promising technique to realize quantum-dot lasers. Since the quantum dots appear spontaneously during growth, they are said to be self-assembling, self-ordered or self-organized. The most important benefits of such an epitaxial self-organization assembly are: a giant number of nanostructures is formed in one simple deposition step; the synthesized nanostructures can reveal a high uniformity in size and composition; they may be covered epitaxially by host material with minimization of crystal or interface defects.

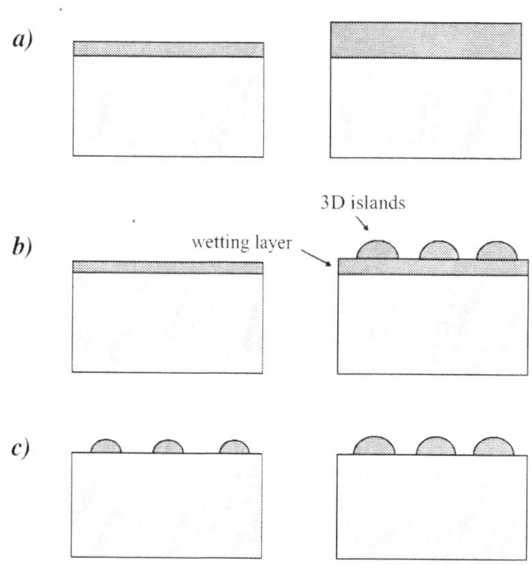

Figure 1. Schematic diagrams of the growth modes for semiconductor systems: *a)* Frank-van der Merwe (FM, layer-by-layer); *b*) Stranski-Krastanow (SK, islands with wetting layer); *c*) Volmer-Weber (VW, islands).

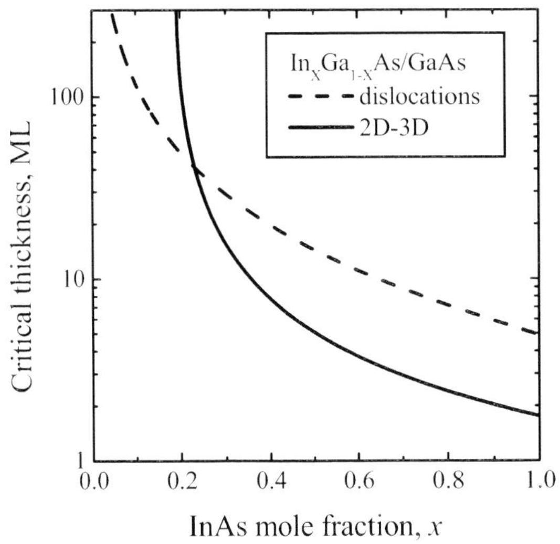

Figure 2. Critical layer thickness as a function of the In composition in InGaAs for the misfit dislocation formation (dashed line) and island growth mode (solid line).

Self-organization of three-dimensional islands has been observed for the first time in InAs/GaAs material system as a result of attempts to grow a short period InAs/GaAs superlattice on a GaAs substrate [5]. Though arrays of self-organized QDs have been synthesized and studied in a number of systems, In(Ga)As/GaAs material system has remained the most intensively investigated and practically exploited.

MBE machine equipped with a high-energy electron diffraction (RHEED) system offers a method of the surface morphology monitoring. Formation of an array of three-dimensional islands is clearly detectable by change in RHEED pattern from streaky to spotty. The driving force for the self-organization process as well as for the misfit dislocation formation is the lattice mismatch between the epitaxial layer and the substrate (buffer layer). In the case of the InAs deposition on Ga(Al)As, the difference in lattice constants reaches 7%. Figure 2 shows the dependence of the 2D-3D critical thickness on the indium mole fraction in the epitaxial layer. The critical thickness for the 60^0-dislocation formation calculated according to the Matthews and Blakeslee theory of the mechanical equilibrium [6] is also added for a comparison.

It is seen that while weakly strained InGaAs layers ($x \leq 0.2$) predominantly accommodate lattice mismatch by misfit dislocations, highly strained InGaAs films ($x \geq 0.5$) grow in a three-dimensional mode prior to the dislocation formation. The critical coverage corresponding to the onset of the island mode is 1.7 ML for the growth of InAs on the GaAs(100) surface [7] and increases with decreasing the lattice mismatch (InAs mole fraction in the epilayer) to 4 ML for the 50% content of InAs [8].

Extensive transmission electron microscopy (TEM) studies have shown that ensemble of InAs quantum dots in a GaAs matrix usually has surface density around 5×10^{10} cm^{-2}, individual island shape is a pyramid with square base [9]. The lateral base size is about 10-15 nm, the height is 3-5 nm. Schematic drawing and plan-view TEM image of InAs/GaAs QDs are shown in Figure 3.

a) *b)*

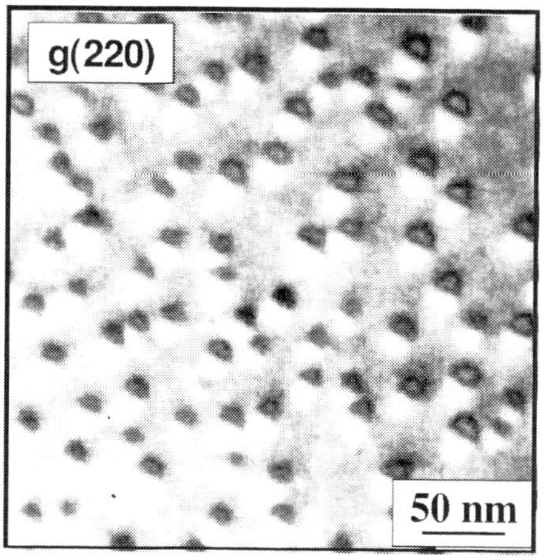

Figure 3. Schematic geometry (a) and plan-view TEM image (b) of self-organized QD array.

As the bandgap energy of InAs is smaller than GaAs one, InAs islands work as potential traps for charge carriers. However, it took several years to demonstrate that these islands can be coherently strained, i.e., free of dislocations [10] and, as a result, demonstrate strong luminescence [11]. It was found that the key factor in the creation of defect-free InAs islands is the precise control of the effective thickness of In(Ga)As deposited, $Q_{In(Ga)As}$ [2]. The results obtained show that optically active islands can be formed within a very limited range of the effective thickness (from 1.7 to about 4 ML for InAs/GaAs).

The peak position of the PL line from an individual QD or QD array is primarily governed by the island size or the average island size, respectively. Under the same deposition condition, the QD size increases with increasing the effective thickness, Q_{InAs}, whereas the QD density steeply increases at the very beginning of three-dimensional growth mode (1.7÷2 ML) and then saturates [12]. This is illustrated in Figure 4, where plan-view TEM images of QD structures grown with 2 or 4 ML of InAs are presented. Significant growth of the islands in size is detectable in the sample with higher Q_{InAs}.

a) *b)*

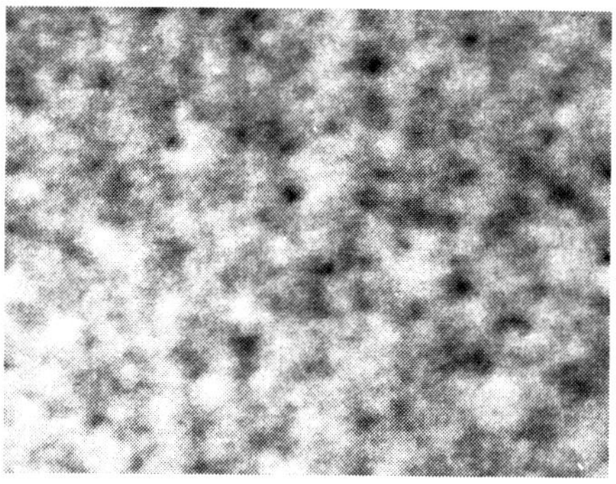

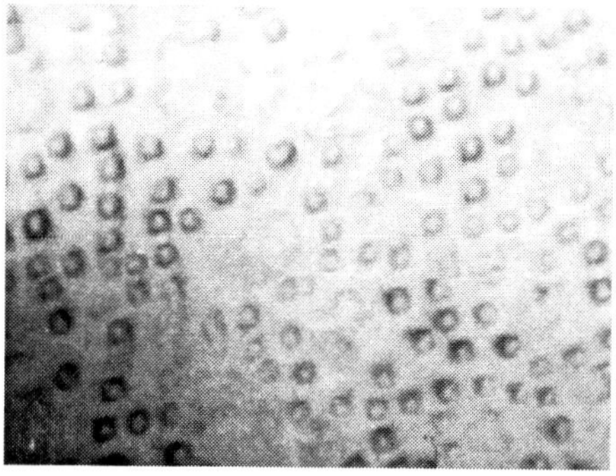

Figure 4. Plan-view TEM images from InAs QDs deposited with QInAs=2 ML (a) or 4 ML (b).

The change in size corresponds to an energy shift of the emitted light, which is due to the altered carrier confinement and strain in the dots. Larger dots emit photons of lower energy, corresponding to longer wavelengths. Room temperature PL spectra of several InAs/GaAs QD samples with different Q_{InAs} are shown in Figure 5,a.

Formation of three-dimensional InAs islands results in the appearance of the intensive band consisting of one or two lines strongly shifted to longer wavelengths with respect to the GaAs band-edge luminescence. The longer wavelength line originates from the radiative recombination of carriers localized in the ground state of QDs. The shorter wavelength line is due to the excited state that appears in samples with higher Q_{InAs} indicating that the island volume becomes large enough to localize the excited level.

Typical linewidth of these PL lines is 40÷60 meV indicating the size fluctuations in the QD ensemble. However, each individual QD is characterized by extremely narrow (less than kT) luminescence line centered at the photon energy, which corresponds to a certain QD size. The splitting of the QD emission band into a series of ultra-narrow lines with decreasing the number of pumped QDs has been demonstrated by high-resolution e-beam excitation [13].

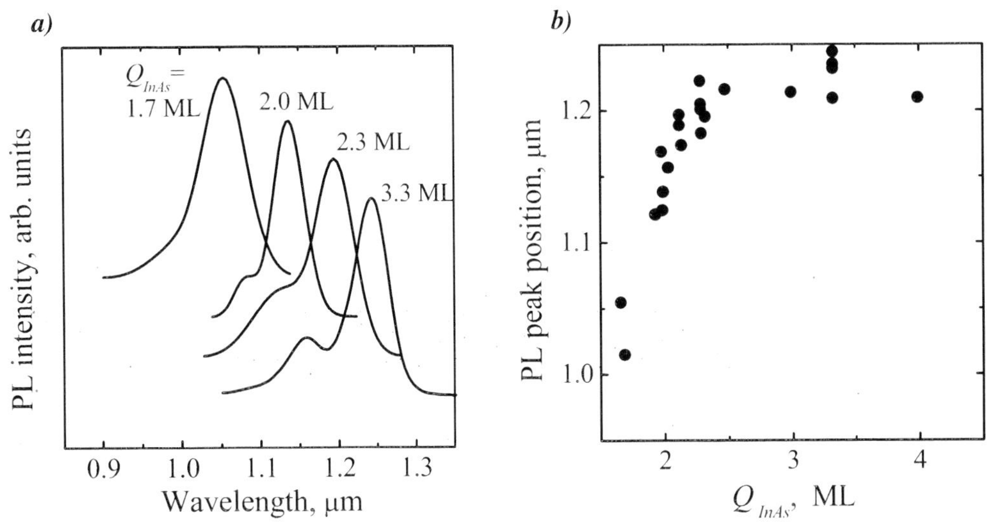

Figure 5. Room temperature PL spectra (a) and peak position (b) of the structures containing InAs quantum dots in a GaAs matrix as functions of the effective thickness of InAs deposited, Q_{InAs}.

Figure 5,b shows the dependence of PL peak position on the thickness of InAs deposited for QD formation. Within the interval of the dislocation-free growth, the QD luminescence covers the wavelengths from 1.05 to 1.24 μm. Note, that InGaAs/GaAs strained QW usually emit not longer than 1.1 μm because of the inherent limitation of the In composition and the QW width due to the plastic strain relaxation. Employing special growth technique (low-temperature growth, strain compensation) one is capable to suppress the growth transition from 2D to 3D growth mode thereby increasing the longer wavelength border of the QW luminescence. For example, Kudo and Mishima have reported a peak PL wavelength of 1.223 μm in the 21-ML thick QW with the high In content of 42% grown by MBE [14]. It should be noted that these well widths exceed the critical thickness predicted by the Matthews and Blakeslee model [6].

It is also worth mentioning that the emission wavelength of QD luminescence saturates with the deposition of the strained material in the excess of 3÷3.5 ML, Figure 5,b. At this coverage the most part of QDs reaches their equilibrium size while further growth results in formation of misfit dislocations.

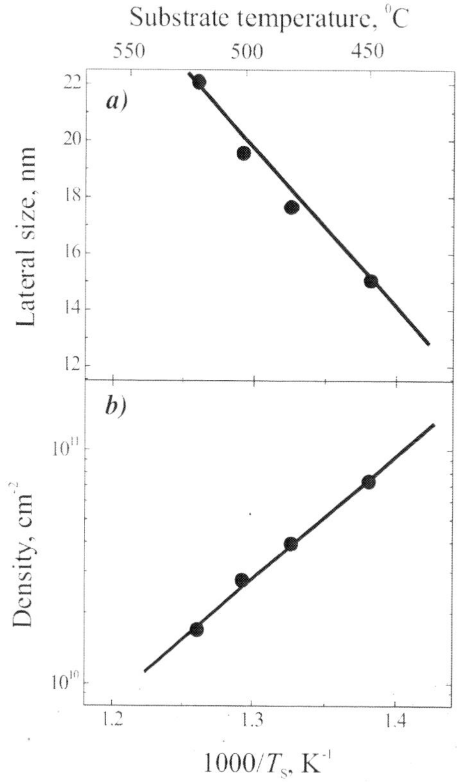

Figure 6. Effect of deposition temperature on average lateral size of InAs/GaAs islands (a) and island density (b).

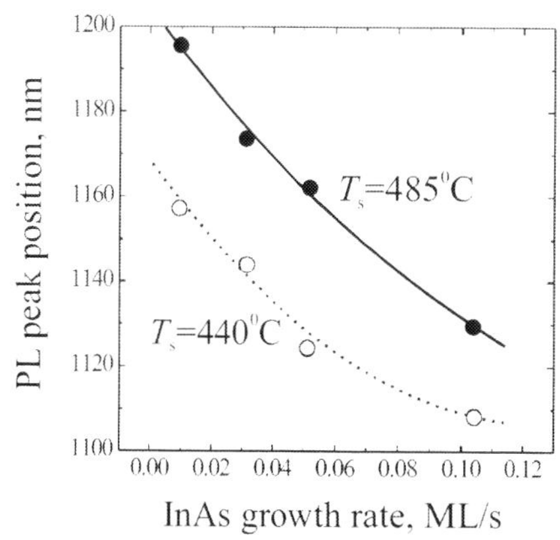

Figure 7. PL peak position of 2ML-InAs QDs as a function of InAs growth rate at different growth temperature (open circles: 440^{0}C, solid circles: 485^{0}C).

Varying growth temperature, InAs deposition rate, and other growth regimes during deposition of QDs allows one to influence on QDs characteristics as like as surface density, QDs size and shape [15, 16]. Temperature is one of the most powerful among growth parameters to vary QDs properties. QDs density and average lateral size demonstrate opposite behavior (see Figure 6): increase of substrate temperature leads to increase of the lateral size of QDs and to decrease of the surface density. Also, it was found that surface density of QDs array deposited at 440°C increases from 3.5×10^{10} to 1.2×10^{11} cm^{-2} if InAs deposition rate changes from 0.01 to 0.1 ML per second; and from 7×10^{9} to 3×10^{10} cm^{-2} at substrate temperature of 485°C within the same range of InAs deposition rate. At the same time average lateral size of QDs decreases from 12 to 10 nm or from 21 to 13 nm, respectively. As a result, PL peak of QD array shifts to shorter wavelength with increasing the growth rate or decreasing the deposition temperature, see Figure 7.

The emission wavelength longer than 1.24 μm can hardly be achieved by bare deposition of In(Ga)As in Strainski-Krastanow growth mode. A specific method of QD deposition had to be introduced to prolong the dislocation-free growth mode and extend the longer wavelength border of the QD luminescence. In 1994 Mukai et al have reported on MOVPE growth of QD structures in a GaAs matrix with the ground state emission centered at 1.32 μm [17]. The authors have pointed out that the atomic layer epitaxy method used, i.e., the alternating deposition of the elements, is better for the layer-by-layer growth mode than conventional MBE or MOVPE. Even if the three-dimensional islands of the group III atoms are formed on the surface, the islands rearrange into a two-dimensional structure when arsine is supplied to the surface. This allowed them to extend the dislocation-free growth mode and to form large QDs (20 nm in base, 10 nm in height) by depositing 24 ML of $In_{0.5}Ga_{0.5}As$. The similar method was later realized in MBE by Mirin et al [18].

A very promising feature of these long-wavelength emitting QDs is a good size uniformity. While the conventional InGaAs self-organized QDs are usually characterized by relatively wide spectral lines, with the typical full width at half maximum of about 60-70 meV, FWHM as low as 33 meV was reported in [19]. The larger migration distance due to the alternating sub-ML deposition cycles yields larger QD size with improved uniformity. On the other hand, atomic force microscopy shows that the 1.3 μm wavelength QDs are formed with a relatively low density of ~ 1.3×10^{10} cm^{-2}. It should be mentioned that the surface density of QD array is one of the key factors affecting the device performance of a QD diode laser [20]. The finite surface density restricts the optical gain which can be achieved at the ground-state transition. It was shown that there is some optimal density of QD array corresponding to the minimal threshold current density for given optical loss loss. If the loss is too high to be overcome at the ground level one should increase the QD surface density, otherwise lasing proceeds via the excited level (λ~1.15 μm).

Limited optical gain of QD lasers with ALE-grown QDs associated with their low surface density motivates a search for novel QD materials suitable for 1.3-μm application. An alternative approach for the formation of long-wavelength emitters on a GaAs substrate utilizes the sensitivity of the energy of optical transition in a quantum dot to the variation of the bandgap of the surrounding matrix. In particular, it was shown that the PL line shifts to the shorter wavelengths with adding Al to the matrix material (replacing a GaAs matrix by the wider bandgap AlGaAs ternary alloy) [21]. This effect is due to the perturbation of the QD energy levels by a certain penetration of wavefunctions into the barriers. In the similar manner, the quantization energy can be decreased by placing the QDs in a narrower bandgap material.

The method is based on capping the self-organized InAs QDs with thin InGaAs layer. It was proposed independently and nearly simultaneously by several research groups [22-24]. An external InGaAs quantum well acts as this narrow bandgap host. Figure 8 shows an evolution of the PL spectrum from a QD array covered by 4-nm-thick InGaAs QW. It is seen that the decrease in the bandgap of the surrounding material strongly red shifts the wavelength of the QD line up to 1.3 μm at room temperature.

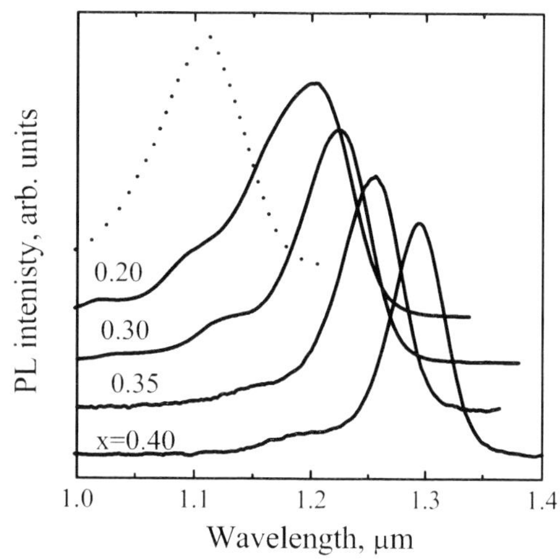

Figure 8. 300K PL spectra of structures containing InAs QDs placed in 4-nm-thick $In_XGa_{1-X}As$ quantum well as functions of the InAs mole fraction in the QW. Dotted line is for InAs QDs in a GaAs matrix.

An alternative explanation for the observed red-shift of the luminescence line is based on observations reported in Ref. [25]. It was found that the emission wavelength of uncapped, i.e., free-standing on GaAs surface, self-organized quantum dots is as long as 1.53 μm. On the other hand, buried QDs usually have an emission wavelength of 1-1.2 μm. This indicates that there is strain relaxation in the non-buried QDs and it is possible to produce long-wavelength optical devices on GaAs substrate through the use of a low-strained dot structure. This idea can be realized by covering the self-organized InAs QDs by a "strain-reducing layer" composed of less-strained material, for example InGaAs or InAlAs for InAs QDs on GaAs. It was shown that the emission wavelength can be extended up to 1.35 μm at room temperature by capping 2.4-ML InAs with a 7-nm-thick $In_{0.2}Ga_{0.8}As$ layer [22]. In the strain-reduced QDs, as the lattice constant of the strain-reducing layer is larger than that of GaAs, lattice constraint in the growth direction of InAs QDs can be relaxed. Then a strain component along the growth direction is partly reduced, resulting in a smaller bandgap shift induced by strain.

Moreover, it was shown that the effect of "stimulated decomposition" of the capping ternary alloy (InGaAs or InAlAs) should be taken into account [26]. When a QD array is overgrown by a ternary alloy, the In atoms of the alloy are distributed nonuniformly in the growth plane owing to the QD induced strain fields. Indium atoms tend to reach such local areas, where the strain is partly relaxed. Thus, indium content of the InGaAs quantum well near and above the island is higher than the average indium composition of the material deposited. As a result, local In-rich regions are formed in the vicinity of the QDs. These are clearly observed in cross-section TEM image of QD array overgrown by InGaAs QW, Figure 9. Thus, the effective size of resultant islands is larger than that of QDs covered by GaAs.

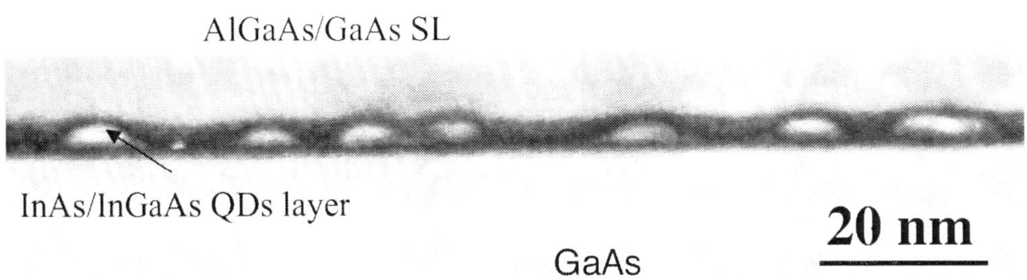

Figure 9. Cross-section TEM images of the sample containing single plane of InAs QDs capped by 10-nm-thick $In_{0.12}Ga_{0.88}As$ QW.

Up to now there is no decisive argument to clarify unambiguously what the explanation, i.e. strain reduction, matrix bandgap reduction, or stimulated decomposition is most significant. Probably, all these effects contribute to a certain extent. In particular, strain reduction and stimulated decomposition can explain why the red-shift of PL line is observed even in the case of capping the QDs by InAlAs or InAlGaAs instead of InGaAs [22, 26], as it is observed in Figure 10. Note that InAlGaAs has significantly larger bandgap energy as compared to InGaAs. Nevertheless the PL peak position of QD luminescence is quite similar.

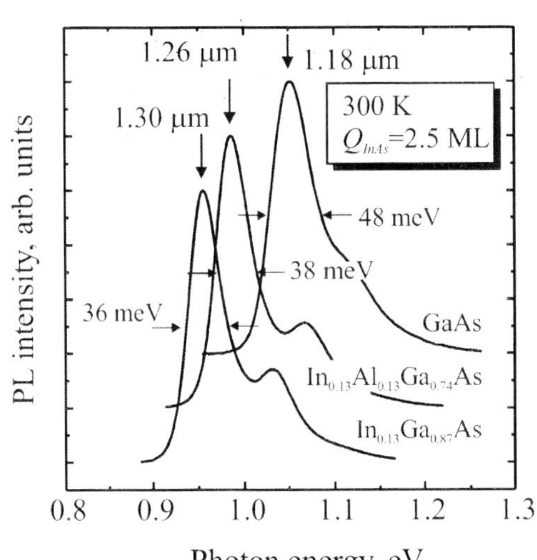

Figure 10. Room temperature PL spectra of InAs QDs (Q_{InAs}=2.5) capped with different materials (GaAs, InGaAs, or InGaAlAs).

It was proposed by L. F. Lester to call these long-wavelength QDs and the method itself as DWELL (dots in well). We believe that the main advantage of this approach, also known as QD-in-QW, over the ALE is that the surface density of quantum dot is as high as 4-5×10^{10} cm^{-2} per QD plane or even higher (~7×10^{10} cm^{-2} [24]). This corresponds well to the usual density of self-organized QDs in a GaAs matrix. Consequently, lasing from a single layer of QDs is possible at reasonable cavity losses.

The wavelength of the QD emission as a function of Q_{InAs} for several compositions of the external $In_XGa_{1-X}As$ QW is presented in Figure 11,a. It is seen that the PL peak position can reach 1.3 μm at certain QW parameters (InAs mole fraction, x, and the thickness, L_Z, for the given Q_{InAs}). Thus, the wavelength of the QD emission can be effectively controlled by the variation of both the effective band-gap of surrounding matrix and the size of QDs themselves. However, the total amount of InAs deposited, $Q_{\Sigma} = Q_{InAs} + x \times L_Z$, is different for different sets of parameters as the wavelength is fixed. Because, both the QDs and the QW are lattice mismatched with respect to GaAs host material there is a risk of misfit dislocation formation. It was found that the minimization of the total In amount, Q_{Σ}, is the most essential for controlling the misfit dislocation formation and achieving bright long-wavelength luminescence [27].

If small QDs are used ($Q_{InAs} \leq 2$ ML) the InAs content in the QW is relatively high and significantly increases Q_{Σ}. On the other hand Q_{InAs} directly contributes to Q_{Σ}. Thus, there should exist an extremum in the dependence of the total amount of InAs, Q_{Σ}, used for the formation of the active region on Q_{InAs}. This dependence is shown in Figure 11,b by open circles for the samples emitting in the 1.29÷1.3 μm range. It does have the minimum at Q_{InAs} around 2.5 ML. In this point the total amount of InAs in the active region is less than 5 ML. This value should be compared with 5.5 ML in ALE-method to achieve the emission at 1.3 μm. It should be noted that the lower values of Q_{Σ} correspond to the higher integrated PL intensity as indicated by solid circles (right axis).

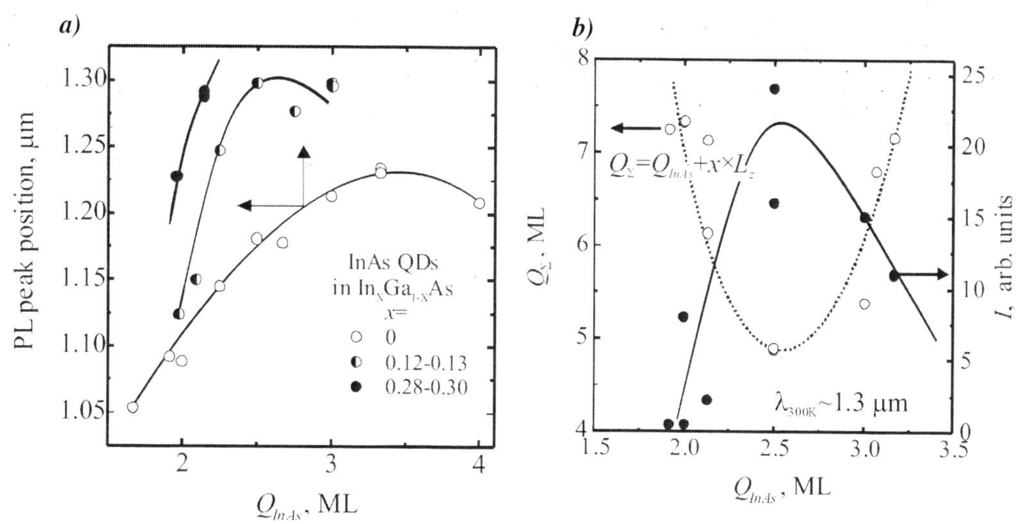

Figure 11. PL peak position at 300K of InAs QD as a function of Q_{InAs} for different QW compositions (a); dependence of the total amount of InAs deposited (QD plus QW), Q_{Σ}, and integrated PL intensity on Q_{InAs}.

Figure 34 summarizes the tunability of the PL line of InAs quantum dots with varying the matrix bandgap. It is seen that the emission wavelength of strained InAs QDs on GaAs can be precisely tuned to the desired spectral range by the properly chosen matrix material thereby covering a wide wavelength from 1.3 to 0.8 μm at room temperature.

3. Pseudomorphic Quantum Dot Lasers

Laser operation in QD heterostructures was demonstrated for the first time in 1994 for the optically pumped structure [28] and shortly after for current-injected diode laser [8, 29]. In these references single plane of self-organized InGaAs QDs inserted in the central part of a GaAs cavity was used as the laser active region. The low-threshold lasing (~100 A/cm^2) on the QD ground state was observed up to 120K with the wavelength of about 1 μm. At higher temperatures, however, the threshold current was increasing up to ~1 kA/cm^2 at room temperature and the lasing wavelength was shifting to the wetting layer transition. This behavior was attributed to thermally activated escape of carriers from QDs to the surrounding layers.

Soon a significant progress in room-temperature characteristics of QD lasers has been done owing to development of multiple stacking of several QD planes [30] in combination with wider bandgap matrix material (AlGaAs) [31]. Ground-state QD lasing with the room temperature threshold current density of 98 A/cm^2 has been reported [32, 33]. Also, it was found that lasers based on multiply stacked QDs offer much better efficiency characteristics [34]. CW operation of a QD laser (for the first time demonstrated in 1996 with the maximum output power of 300 mW [33]) has been shortly after improved to 1W by wider bandgap matrix [35].

However, the above-mentioned QD lasers operated in the spectral range around 1 μm. Room temperature lasing at the wavelength of 1.3 μm was achieved for the first time in 1998 from the ground state of the ALE-grown QD ensemble [36]. Ground state lasing was achieved up to 295-K with the threshold current density of 405 A/cm^2 and lasing wavelength of 1.31 μm. A jump to the first excited state was apparent in the lasing spectra between 295 and 300 K in this as-cleaved laser. Moreover, the above-threshold differential quantum efficiency was as low as 3%.

Both these facts indicate that the saturated gain is extremely low. The low QD density that provides the low transparency current density also results in a small maximum (saturated) gain. The authors of Ref. [36] pointed out that this type of QD laser is best suited for low loss cavity designs that can utilize the low transparency current. HR/HR facet coatings provide the lowest possible cavity loss owing to the minimization of the mirror loss. Using this approach, the region of the ground state lasing was extended up to 324 K with J_{th} of 270 A/cm^2 at 300 K.

Afterward the record-low threshold current density among ALE-grown QD lasers of 19A/cm^2 has been reported in [37]. The lasing wavelength was 1.33 μm. The key point was to deposit thin $In_{0.09}Ga_{0.91}As$ quantum well before QD planes. The authors found that the thin InGaAs buffer layer increases the QD density by nearly a factor of 2 as compared to when a GaAs buffer layer is used (from 1.2×10^{10} to ~2×10^{10} cm^{-2}). The authors pointed that the increase in dot density contributes to the low threshold CW operation for the devices of that work. Also it should be noted that the use of the oxide-confined waveguide allowed authors to prevent lateral current spreading thereby reducing the threshold current. However, the efficiency and power characteristic were unacceptably low: ~0.02 W/A with the maximum output power in CW as low as 300 μW. The authors themselves pointed to the cavity design (99%-reflective facet coatings) and the QD material quality as possible reasons of the low output power and efficiency for these 1.3 μm QD lasers.

A tradeoff between threshold and efficiency characteristics is among the main issues of QD laser design. We believe that the main disadvantage of the above ALE approach is associated with the mentioned low density of QDs in the ensemble. The QD density decreases from $\sim 5\times10^{10}$ cm^{-2}, which is typical for self-organized InGaAs QDs, to $(1.0\text{-}1.5)\times10^{10}$ cm^{-2} when ALE growth mode with pauses under As flux is used [19]. The QDs formed in this way are essentially larger and, consequently, demonstrate longer wavelength as compared to the conventional QDs (~50 nm vs ~35 nm according to AFM) at the expense of the lower density. This originates directly from the larger migration distance, which is inherent of the migration-enhanced epitaxy. Low QD density and resultant low maximum gain restricts possible cavity design to long cavities or HR/HR coatings. Such low-loss design makes it difficult to achieve simultaneously high differential efficiency and low threshold current.

The fact that the optical gain of a QD laser can not exceed some maximum value, the so-called saturated gain, is associated with finite number of available electronic states in the QD active region. The saturated gain, G^{sat}, of a QD laser is proportional to the density of states at its maximum:

$$G^{sat} \propto g n_{QD} / \Delta \qquad (1),$$

where g is the degeneracy of a QD level, n_{QD} is the QD surface density, Δ is the energy width of density of states for QD ensemble. For the ideal QD array, Δ is given by the homogeneous broadening. In this case, Δ of the order of 0.1 meV or less is expected. However, the energy level of the QD ensemble formed by a self-organization process is additionally inhomogeneously broadened. This proportionally decreases G^{sat}. Thus, the gain saturation problem can be much more serious for real lasers based on self-organized QDs as compared to the ideal array

Figure 12 shows experimental correlation between threshold modal gain (G_{th}), lasing wavelength (λ) and threshold current density (J_{th}) measured on 1.3-μm QD lasers based on three planes of InAs QDs in InGaAs quantum wells. Optical loss (threshold gain) was varied by changing the cavity length. Two distinct regions are clearly observed in the gain-current curve. In each region optical gain saturates at a certain level, which depends on the quantum level index. It should be noted that the onset of the second region in the curve corresponds to the abrupt blue shift of the lasing wavelength. Therefore, this behavior was attributed to the ground-to-excited state transition.

Very similar behavior was experimentally observed in QD lasers with the active region based on InGaAs QDs in an AlGaAs matrix (λ ~ 0.85 μm) [38] and InAs QDs in an InGaAs matrix on InP substrate (λ ~ 1.8 μm) [39]. This finding allows us to make a conclusion that the gain saturation and ground-to-excited state transition is a general feature of any diode laser based on self-organized QDs.

Closer look to correlation between lasing wavelength and optical loss is presented in Figure 13. PL spectrum of QD array is superimposed to this dependence. For this example we use data of QD-in-QW laser with 10 planes of QDs. It is seen that lasing line (1.29 μm at the lowest loss) is shifted to longer wavelengths with respect to the maximum of PL peak of QD ground state (1.25 μm in case of Figure 13). With increasing optical loss, lasing wavelength is gradually blue-shifted getting closer to PL maximum. However, the wavelength changes much slowly within the region of ground-state lasing as compared to ground-to-excited state lasing transition, because in the former case lasing proceeds on the same inhomogeneously broadened level of QD ensemble.

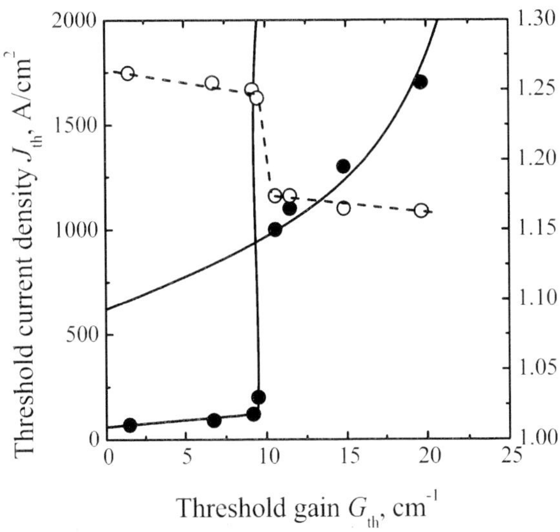

Figure 12. Experimental dependence of the threshold current density, J_{th} (solid circles) and lasing wavelength λ (open circles) on the threshold gain, G_{th} for QD laser based on three planes of InAs/InGaAs quantum dots. Solid lines are calculated J_{th}-G_{th} dependence.

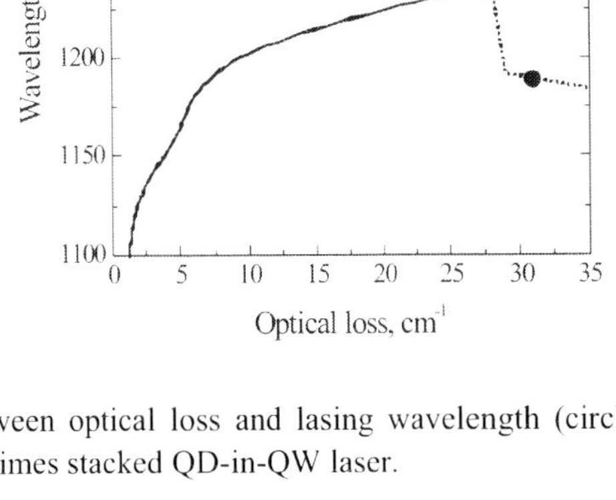

Figure 13. Correlation between optical loss and lasing wavelength (circles and dotted line) and PL spectrum (solid line) of 10-times stacked QD-in-QW laser.

Simple analytical expression that describes dependence of the QD optical gain on pump current has been evolved in Ref. [40, 20]. Using properly chosen gain parameters, such as saturated gain G^{sat} as well as the transparency current density and its components, it is possible to fit the gain-current density characteristic for both ground- and excited-state lasing with good accuracy, as shown in Figure 12. Calculated results are shown by solid lines. For the particular case of QD laser formed by three QD-in-QW planes shown in the figure, the best suitable fitting parameters for the ground-sate saturated gain is G^{sat}=9.5 cm^{-1}. Thus, the value of saturated gain per one plane of self-organized QDs with typical density of about 5×10^{10} cm^{-2} is usually about 3-4 cm^{-1}. Because the saturated gain is a linear function of the surface density of QD array, a ratio G^{sat}/n_{QD} is among the intrinsic parameters of QD array. Of course, it depends on the optical confinement of the laser waveguide. However, the latter value can hardly be changed in a wide range (orders of magnitude). The G^{sat}/n_{QD} ratio was estimated to be $\sim1\times10^{-10}$ cm^{-1}/cm^{-2}.

Since the optical gain achieved on the ground subband of a QD array is limited by its saturated value, the ground state lasing is only possible if the total optical loss (internal plus mirror) does not exceed the saturated gain. Transition to excited state lasing is accompanied by an abrupt blue shift of the lasing wavelength and significant increase in the threshold current density due to higher transparency current density of the excited subband. To keep operating wavelength of a QD laser near 1.3-μm and prevent steep superlinear increase of the threshold current density in the regime of high output loss, the saturated gain of the ground subband should significantly exceed the necessary level of the optical loss.

The saturated gain can be improved by use of denser QD arrays or multiple stacking of several QD planes. In the latter case n_{QD} in Eq. (1) plays a role of the total density of QDs in the laser active region (density of QDs in a single plane multiplied by the number of QD planes). It was shown previously that the ALE-grown InGaAs QDs can be stacked at least twice [19]. On the other hand, QD ensemble formed by QD-in-QW method looks more preferable because of their higher density as compared to that of ALE-formed QDs. From this viewpoint, already single plane of QD-in-QW is nearly equivalent to triply-stacked QD array grown by ALE method.

The transparency current density is also proportional to the total surface density of QD array:

$$J_0 = \frac{qn_{QD}}{\tau} \qquad (2),$$

where q is the electron's charge and τ is the recombination lifetime. Thus, the increase in the QD surface density results not only in a corresponding growth of the saturated gain but the transparency current density as well. The competition between the transparency current density and the gain saturation determines the threshold current density at the given cavity loss. Therefore, there should be some optimum surface density of QDs (optimum number of QD planes) which ensures the minimum threshold current density for a given optical loss.

Figure 14,a shows the calculated dependence of the threshold current density on the total surface density of QD array in the active region with total optical loss as a parameter [40]. Gain parameters were extracted from the experimental data on 1.3 μm lasers based on InAs/InGaAs QDs (Figure 12). It is seen that the plotted dependence is qualitatively very similar to the dependence of the threshold current density on the number of quantum wells in a QW laser. J_{th} of a QD laser initially decreases with decreasing the surface density of the array due to the corresponding reduction of the transparency current density (shown by dashed line). There exists some optimal QD density, n_{QD}^{opt}, corresponding to the minimal threshold current density, J_{th}^{min}, for the given total loss, α. Further decrease of the surface density below the optimum value leads to the rapid increase of the threshold current density as a result of the gain saturation on the ground state. If the necessary gain is too high for the given QD surface density and exceeds the saturated gain the ground-state lasing is never achieved at any current density. However, lasing may be possible through the excited subband.

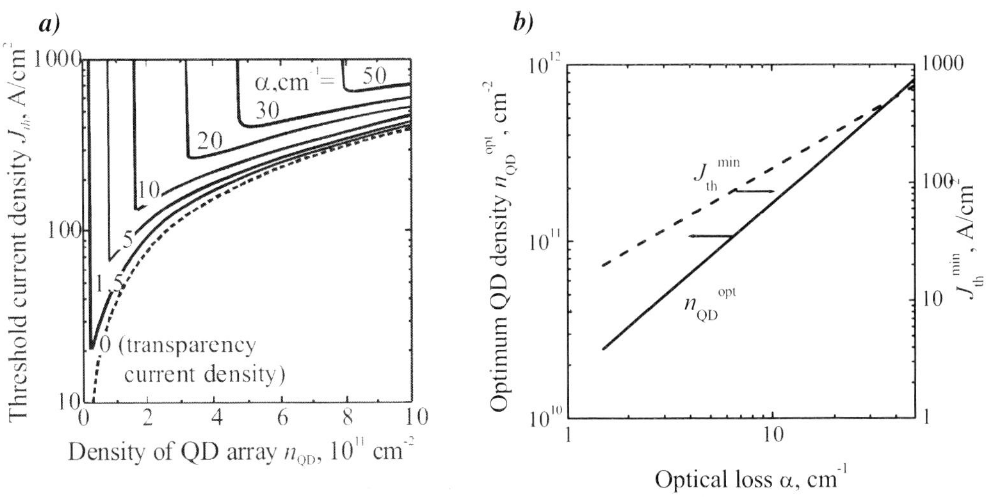

Figure 14. (*a*) Calculated dependence of the threshold current density, J_{th}, on surface density of QD array, n_{QD}, at different optical loss, α, with gain parameters extracted from data of Figure 12; (*b*) Dependence of the minimum threshold current density, J_{th}^{min}, and corresponding optimum surface density of QD array, n_{QD}^{opt}, on optical loss, α.

The optimal density of the QD array and the corresponding minimal threshold current density are shown in Figure 14,b as a function of the total loss. It is seen that both n_{QD}^{opt} and J_{th}^{min} can be well approximated by linear functions. For most of device applications the required optical loss is relatively high (more than 10 cm^{-1}) since shorter laser cavity can provide higher external differential efficiency. Data presented in Figure 14,b show that the optimal surface density for that level of the loss exceeds 10^{11} cm^{-2}, which is several times as much as typical surface density of a single self-organized QD plane. Consequently, denser QD arrays deposited by special technological methods or multiple stacking are most suitable for laser applications.

If several QD planes are stacked in the laser waveguide region the side QD planes are shifted out of the maximum of optical field. Thus, the contribution of additionally stacked QD planes to the total optical gain is reduced as compared to that of the central planes, i.e., the average optical confinement factor decreases with increasing the number of QD planes. Saturated gain of the multiple stacked QD array is shown in Figure 15 as a function of the QD planes number for 0.4-μm-wide SCH region and 30-nm-wide spacer layers between QDs.

The Γ-factor compression associated with the effect of side planes results in sub-linear dependence of the saturated gain on the number of QD planes. This effect is more pronounced in the case of large number of QD planes (more than 10). From this viewpoint it would be advisable to use thinner spacer layers. This way is however in contradiction with plastic strain relaxation due to mutual effect of closely spaced QD planes. More suitable is to increase the QD density in each plane because this method would provide location of the larger number of quantum dots near the mode maximum. Moreover, it would result in the reduction of the wetting layer effect on the threshold current density.

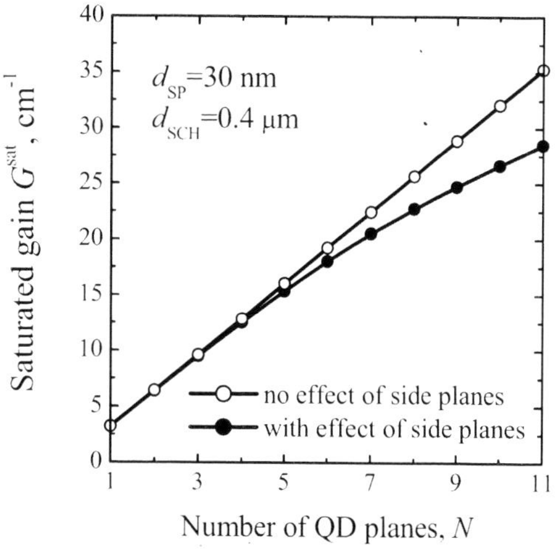

Figure 15. Saturated gain on QD ground state, G^{sat}, as a function of number of QD planes, N. Saturated gain of one QD plane is assumed to be 3.2 cm^{-1}.

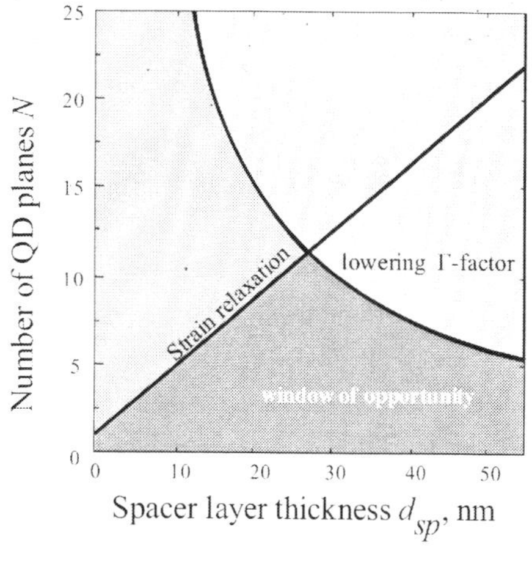

Figure 16. Effect of the spacer thickness on possible number of QD-in-QW planes due to strain relaxation and lowering the optical confinement factor in case of thin and thick spacers, respectively.

Number of QD planes and spacer thickness correlate each other and can not be arbitrarily chosen. Two opposite effects set the range of their possible variation as illustrated in Figure 16. In addition to mentioned effect of side planes (lowering the optical confinement factor), strain relaxation in the multiply-stacked QD region as a whole should be taken into account.

More number of QD planes requires thicker spacers to prevent formation of dislocations. For the 1.3-μm QD-in-QW region up to ten planes can be stacked without any evidence of dislocation formation if the spacer thickness exceeds 30 nm. At the same time, the compression of optical gain is only 25% as shown in Figure 15, and the saturated gain as high as ~25 cm^{-1} can be achieved.

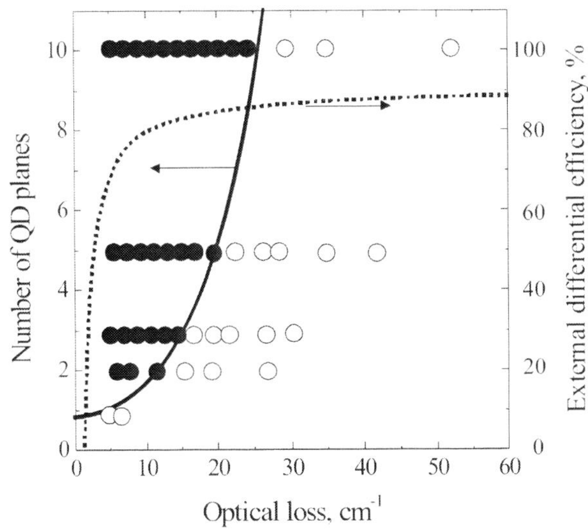

Figure 17. Correlation between optical loss in the laser cavity and number of QD planes which provides ground-state lasing (solid circles) or excited state lasing (open circles). Solid line separates the regions of ground-state and excited-state lasing. Dotted line shows the calculated dependence of external differential efficiency on optical loss.

Figure 17 summarizes our experimental data on lasing wavelengths of the QD-in-QW lasers with different number of QD planes as a function of optical loss in the laser cavity. Solid and open circles correspond to laser diodes operating at either ground-state (~1.3 μm) or excited-state (~1.17 μm) transition, respectively. Solid line in Figure 17 represents the calculated number of QD-in-QW planes, n_{QD}^{min}, which is necessary to deposit in the active region to achieve the ground-state lasing at the given level of optical loss. The regions on and above this line correspond to the ground state lasing whereas the region below corresponds to the lasing at the excited states. A good agreement between experimental results and theoretical predictions is clearly seen. The n_{QD} from 5 to 10 offers sufficient gain for the practically exploited optical losses (cavity lengths).

Dotted line in Figure 17 shows the differential quantum efficiency, η_D, as a function of the cavity loss, α, calculated by well-known equation:

$$\eta_D = \eta_i \frac{\alpha - \alpha_i}{\alpha} \tag{3}$$

As a typical case we assume the internal loss, α_i, of 1.2 cm^{-1} and the internal differential efficiency, η_i, of 90%. The benefit of the high-loss regime and higher n_{QD} is clearly indicated by the increase in the differential quantum efficiency.

Very promising efficiency characteristics with maximum η_D of 57% has been demonstrated in triply-stacked QD-in-QW laser [41]. Also, the external differential efficiency in excess of 50% were obtained by the group of the University of New Mexico [42]. However further increase in a number of planes, that results in corresponding increase of the saturated gain, typically led to degradation of threshold current and internal quantum efficiency of 1.3-μm-QD lasers due to point defects caused by too high strain accumulated in the active region.

Very recently a significant progress in efficiency characteristics of long-wavelength QD lasers has been achieved. Careful optimization of growth regimes and design of spacer layers allowed us fabricating lasers with up to 10 QD-in-QW planes in the active region [43]. This enabled to improve significantly the internal differential efficiency (~95%), decrease the internal loss (to 1.2 cm^{-1}), and increase the maximum optical gain (to 24 cm^{-1}) [44]. As the result, the maximum *external* differential efficiency was as high as 88% [45]. To our best knowledge, this is among the best efficiency characteristics ever reported for QD lasers of any spectral range [46, 47].

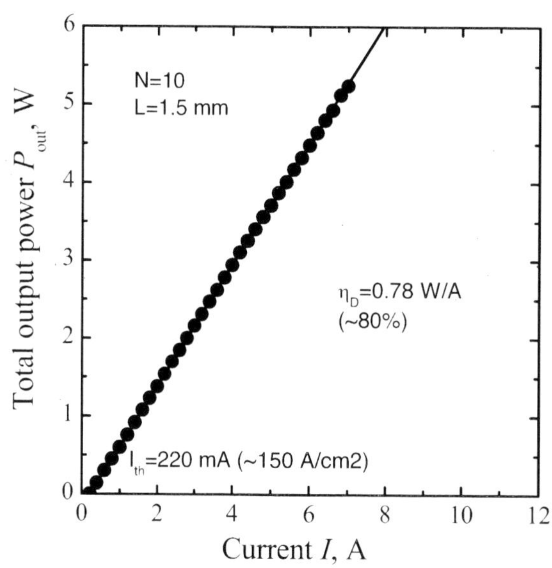

Figure 18. Total output power as a function of current for broad-area 1.5 mm-long laser with active region based on 10 QD-in-QW layers operating at room temperature in pulse regime.

Example of light-current characteristic of representative broad-area QD laser is shown in Figure 18. The output power exceeds 5W without any current-induced transition to excited-state lasing [48]. It is seen that 1.3-μm QD lasers offer a promising combination of low threshold current density and simultaneously high slope efficiency. Dependence of external differential efficiency on the laser cavity length is shown on Figure 19. All over the range of the ground-state lasing the external differential efficiency exceeds 75%. The internal loss (α_i) and the internal differential efficiency (η_i) were then estimated from the experimental $1/\eta_D$-L

dependence. In the particular case shown in Figure 19 values of α_i and η_i were estimated to be 93% and 1.2 cm^{-1}, respectively.

Low internal loss, which is typically of about 1cm^{-1}, is a very attractive feature of QD lasers. As a result the external efficiency is not very sensitive to variation of laser design like cavity length or facet coating: Low internal loss ensures high slope efficiency even in long diodes. It is seen that the maximum value of the slope efficiency is achieved near the shortest cavity length (maximum loss) of the ground state-lasing. In structures based on smaller number of QD-in-QW planes the maximum output loss of the ground state-lasing decreases (the minimum L increases). Correspondingly, the maximum η_D slightly decreases to 84 and 80% (N=5 or 2) due to earlier transition to the excited-state lasing. This highlights the importance of a high saturated gain to achieve highly-efficient QD laser. In particular, a modal gain as high as 24 cm^{-1} was measured in the structure based on 10 QD planes.

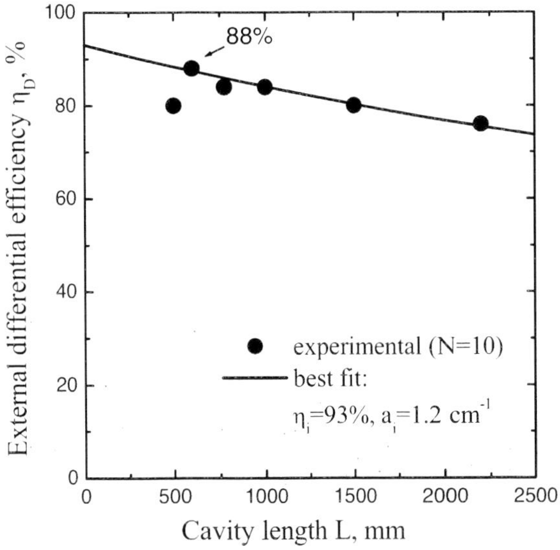

Figure 19. External differential efficiency as a function of cavity length. (close symbol - experiment, curve- calculation)

Internal loss, α_i, and internal differential efficiency, η_i, for the structures with different number of QD-in-QW planes are summarized in Figure 20 as functions of the total surface density of QD array (assuming 5×10^{10} cm^{-2} in each plane). For comparison, the results are also shown, which were measured on laser structures with QDs in a GaAs matrix (no QW cap layer was deposited) and emitting near 1.0-1.1 μm. It is seen that the internal loss regularly decreases as the QD surface density increases. A noticeable decrease of α_i is also observed in QD-in-QW structures as compared to that of their QW-free counterparts. The results obtained can be explained under assumption that the internal loss in QD lasers is mainly contributed by free-carriers in the waveguide region. As the localization of the active region gets deeper (QD-in-QW), the population of GaAs matrix is strongly reduced. Also, higher gain of the multi-plane structure provides inverse population of the active region at lower carrier concentrations thereby decreasing the free-carrier loss.

On the one hand, the internal differential efficiency does not demonstrate a regular dependence on the QD surface density. On the other hand, η_i increases from about 70% to 95% as a result of capping the QD plane by QW. It was shown in [49] that the efficiency of carrier injection into the active region is commonly the main limiting factor for the internal differential efficiency of a QW laser. This can be especially true in case of a QD laser because of a small fill factor of the QDs in the *p-n* junction plane. At the same time, an external quantum well surrounding a QD plane provides an efficient path for carrier capture to the QD states. The importance of the QW role in improving efficiency characteristics was also pointed in [50].

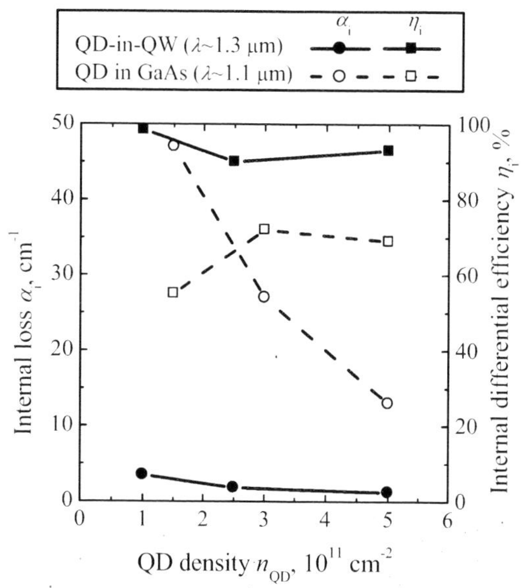

Figure 20. Dependence of internal loss (α_i, circles) and internal differential efficiency (η_i, squares) on total density of quantum dots in the laser active region (n_{QD}) for laser structures of similar design and based on either QDs in QW (solid symbols, solid lines) or QDs in bare GaAs matrix (open symbols, dashed lines).

The transparency current density, J_0, of a laser based on multiply stacked QDs typically increases with increasing the number of QD planes, N. This effect is primarily explained by simple multiplication of total surface density of QDs. In accordance with Eq. (2) J_0 as low as 5 A/cm^2 per QD plane is expected for 1.3-μm QDs formed by QD-in-QW method assuming the radiative recombination in QD ground state with the recombination lifetime τ of the order of 1÷2 ns [51] and taking into account typical surface density, n_{QD}, of about 5×10^{10} cm^{-2}. It is also worth mentioning that multiple stacking of several QD planes may enhance the rate of non-radiative recombination. This may result in superlinear growth of J_0 with N and increase of normalized value J_0/N.

Experimental values of J_0 and J_0/N estimated for those QD-in-QW lasers [43] are shown in Figure 21 by solid and open circles, respectively, as a function of the number of QD planes, N. No evidence of superlinear growth of J_0 with N is observed. Moreover, the transparency current density per QD plane was estimated as 5-7 A/cm^2. These findings prove the absence of noticeable nonradiative recombination although relatively large number of planes was used.

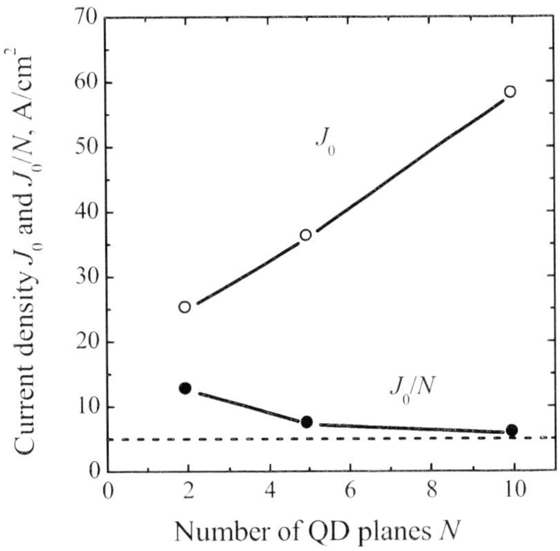

Figure 21. Dependence of total (J_0, open circles) and normalized (J_0/N, solid circles) transparency current density on number of QD planes, N, for laser structures based on multiply-stacked QD-in-QW. The estimated J_0 is shown by dashed line.

QD lasers were also tested in CW regime. CW operation of 1.3-μm QD diode lasers has been reported for the first time in 1998 [52] owing to considerable reduction in the threshold current density along with high slope efficiency. The maximum output power of 2.7 W from 200-μm-wide facet limited by thermal rollover together with the maximum power conversion efficiency of 20% was reported. Significant junction overheating at high drive currents caused by high turn-on voltage (1.5-1.6 V) and high series resistance (>5×10^{-4} □cm^2) resulted in limitation of device parameters.

For a long time these poor diode characteristics were the primary obstacles for further improvement of CW performance. In fact, the inherent reason was associated with the use of high Al content in $Al_XGa_{1-X}As$ cladding layers (x~70-80%), which ensured high optical confinement factor. However, this also led to additional potential barriers and low mobility of charge carriers. Development of multiply stacked QD arrays allowed one to exploit conventional design of GaAs-based diode lasers with $Al_{0.3}Ga_{0.7}As$ claddings. High optical gain minimizes the negative effect of lower optical confinement on threshold characteristics. Also, smaller contrast of the refractive index between claddings and waveguiding layer, which is typically GaAs or $Al_{0.15}Ga_{0.85}As$, ensures smaller beam divergence in vertical direction.

Fig 22 shows the dependence of output power, diode voltage and power conversion efficiency on drive current of 1.3-μm QD laser with improved design [53]. The device's low resistance (~2×10^{-4} Ωcm^2) and low turn-on voltage of 1.1 V along with a differential efficiency of 88% result in total conversion efficiency of 56%. The maximum CW power reaches 5W. To our best knowledge these are the highest values for GaAs-based 1.3 μm lasers.

Reliability of QD lasers was also preliminary tested [54]. A set of four QD lasers was subjected to accelerated aging test at 40^0C heatsink temperature with constant CW drive current of 2A. The output power (about 0.9W) was monitored. During the first 200-300 hours of operation, the output power increased by 50-100 mW due to the initial burn-in process. After 300 hours, the output power stabilized at about 880 mW. No signs of mortality, sudden degradation or wear-out behavior have been observed during 3000 hours and the devices continue to operate. Thus QDs prove their applicability for devices where very high power density is required.

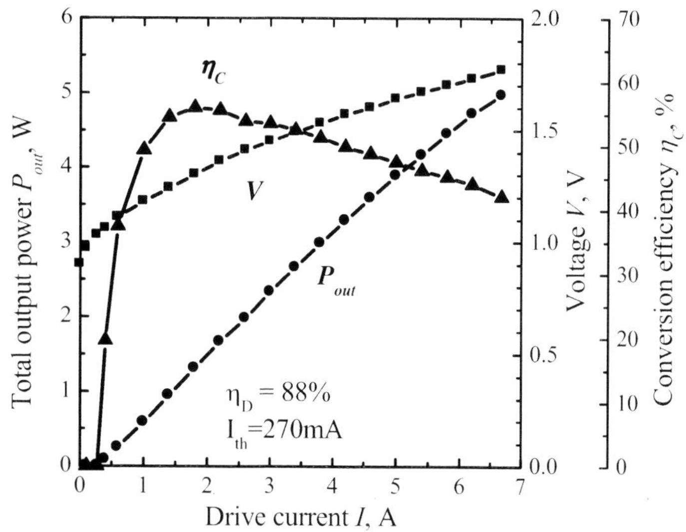

Figure 22. Optical output power, P_{out} (circles), diode voltage, V (squares) and power conversion efficiency, η_C (triangles) as functions of drive current, I, recorded in CW regime at 15^0C for 100μm-wide 1.9mm-long stripe.

Fiber optical communication systems typically require single-mode devices to achieve efficient laser-to-fiber coupling. In case of in-plane (edge-emitting) lasers this means that the laser stripe should be relatively narrow. The problem of the QD laser technology for single-mode operation is associated with additional internal loss caused by the narrow stripe geometry due to light scattering at sidewall imperfections. This loss enhancement can be critical for a QD laser taking into account limited optical gain at the QD ground state. Thus, the realization of a single-mode QD laser is challenged.

Single-mode QD lasers were successfully fabricated by reactive ion etching to form unburred ridge waveguide (RWG) of about 5 μm width [55-57]. Figure 23 shows the light-current characteristics of singlemode QD laser intended for low-power applications. Minimization of the threshold current is one of the most important task in this case because of its significant contribution to the laser operation current. Features of the laser design are short cavity length in combination with high-reflective coatings and use of two QD planes. All these provide extremely low threshold current of 1.4mA ($50A/cm^2$). To our best knowledge this figure is the lowest threshold for any diode lasers with unburied RWG. The output power exceeds 5mW which is sufficient for short-distance communication. Lasing wavelength is exactly 1310 nm at operating current that corresponds to the minimum dispersion of the fiber.

Alternative laser design is intended for high power output. A 2-mm-long cavity is used to improve heat dissipation and diode resistance. Owing to very low internal loss even long cavity can offer high differential efficiency. In this case the slope efficiency exceeds 50%, see Figure 24,a. Up to the maximum drive current the output beam is single-mode as confirmed by far-field pattern shown in Figure 24,b. The maximum power exceeds 300mW. This can be useful to extend the distance of optical communication. The maximum power conversion efficiency is about 30%. The maximum operation temperature of these single-mode QD lasers is as high as 200^0C (sub-kA/cm^2 level of the threshold current density).

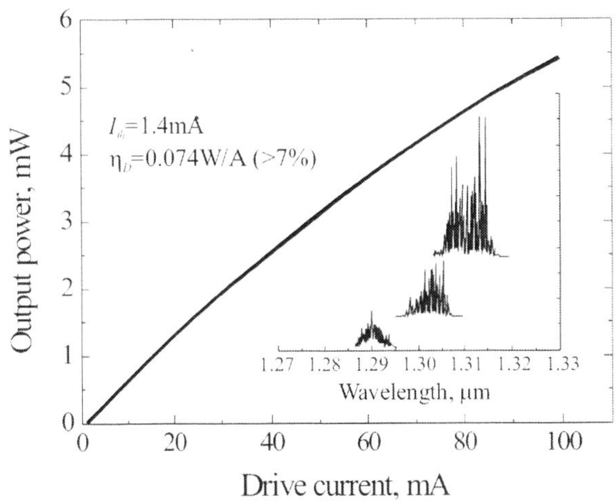

Figure 23. Output power as a function of current (W=4.5μm, L=600μm, HR/HR) at 25^0C. Lasing spectra at different pump currents are shown in the insert.

Temperature dependence of the threshold current is usually characterized by its characteristic temperature T_0 in the equation $I(T)=I_0\exp(T/T_0)$. High characteristic temperature, i.e. weak temperature dependence, is a key factor for potential application of long-wavelength lasers in fiber datalink systems. High T_0 would make it possible to create heatsink-free (uncooled) transmitter modules thereby giving a drastic reduction in cost. One of the main limitations of InP-based 1.3-μm lasers of the present status is low T_0 of 50-70 K [58]. The calculation of Kondow et al [1] shows that this poor characteristics is directly associated with low band discontinuity in the conduction band of InP-based lasers. When the quantum well is so deep that electrons can not overflow to the barrier layer, T_0 is assumed to be 180 K, i.e., the intrinsic value for a QW laser diode.

QD laser potentially offers much better temperature characteristics, because the transparency current density and gain-current characteristic are independent on temperature, see Eq. 1 and 2. In agreement with this prediction already the first QD lasers have demonstrated weak temperature sensitivity of the threshold current or even negative characteristic temperature [59, 60] at low observation temperatures. However, for a long period experimental results on QD lasers demonstrated much worse characteristic temperature, usually around 50÷60 K, near room temperature.

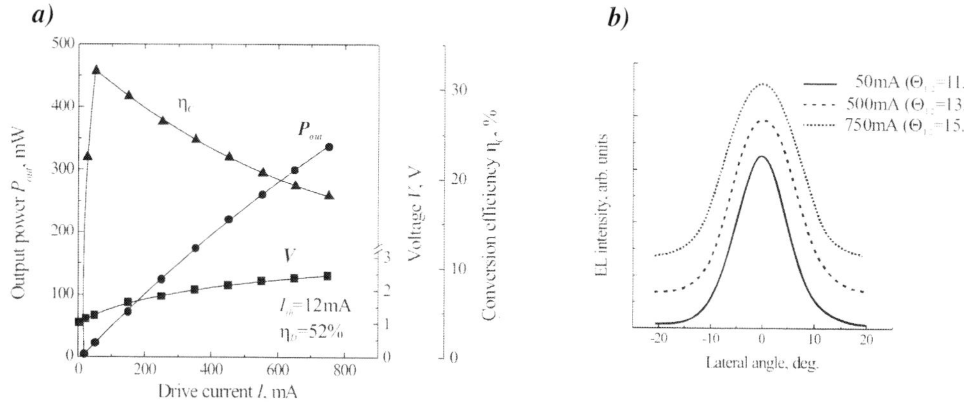

Figure 24. Total output power, voltage and convention efficiency as a function of current (a) and far-field patterns at different current (b) for 1.3-μm QD diode laser (W=4.5μm, L=2050μm, CL/CL facets).

Typical temperature behavior of the threshold current density and lasing wavelength of those QD lasers is presented in Figure 25 (solid symbols). It was found that there is some temperature boundary that separates the regions of high and low values of T_0 while the lasing wavelength exhibits monotonous shift with temperature. A pronounced knee in J_{th}-T curve is clearly observed in Figure 25. Later on it was found that this boundary increases, i.e. the region of high characteristic temperature expands, with increasing the saturated gain and localization energy of the QD active region [41]. Near-room temperature T_0 was also found to be gain dependent [24].

It was suggested that threshold current density at elevated temperatures and its temperature degradation originates mostly from the thermal population of the higher energy states, which also includes the matrix, rather than the QD ground state itself. Development of multiply-stacked QDs as well as long-wavelength QD lasers based on QD-in-QW design lead to noticeable enhancement of the laser gain and deeper localization of the ground state. All these result in significant suppression of the matrix population and consequent improvement of temperature characteristics of QD lasers. This behavior is also in consent with the observed improvement of the internal loss (Figure 19).

In particular, the 1.3-μm QD active region (optical transition at ~950 meV) is characterized by the effective separation of the ground state as high as 450-470 meV with respect to the GaAs bandgap of 1424 meV. At the same time, conventional QDs formed in a GaAs matrix have the emission wavelength of about 1.1 μm, thus the localization does not exceed 300 meV. Example in Figure 25 illustrates how the longer wavelength of QD emission and higher number of QD planes affect the temperature dependence of the threshold current. Very promising results have been achieved in 1.3-μm QD lasers with the active region based on five planes of QD-in-QW [44, 45]. The characteristic temperature was estimated to be 150K up to 80-90^{0}C.

In addition, temperature dependence of a QD laser was found to be affected to a great extent by the energy separation between the discrete levels. However, the hole levels in QDs are very close in energy. The result is that the injected holes are thermally broadened by their closely spaced energy levels. The thermal broadening of the holes decreases the ground state gain and makes the saturated gain quite temperature sensitive. Further progress in temperature characteristics of long-wavelength QD lasers has been achieved by means of p-type modulation doping of the laser active region. The authors from the University of Texas at Austin demonstrated both theoretically [61] and experimentally [62, 63] that doping of QDs with holes can substantially increase the QD laser ground state gain at room temperature and above. In particular, it was shown that when doping increases (0, 26 and 52 holes per QD) the characteristic temperature rises from 85K in undoped sample to 145 and 161 K in p-type doped samples, respectively.

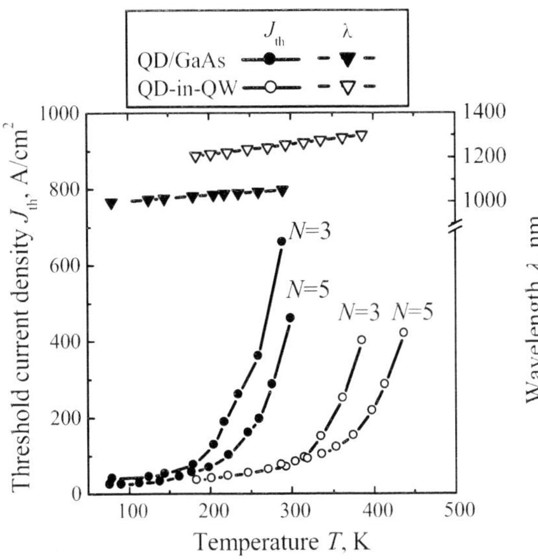

Figure 25. Temperature dependences of the minimum threshold current density (circles) and lasing wavelength (triangles) for QD lasers with different number of QD planes, N, of different design: InAs QDs in as GaAs matrix (λ~1.05 μm) (solid symbls) or InAs QDs capped with InGaAs quantum well (QD-in-QW design, λ~1.27 μm, open symbols).

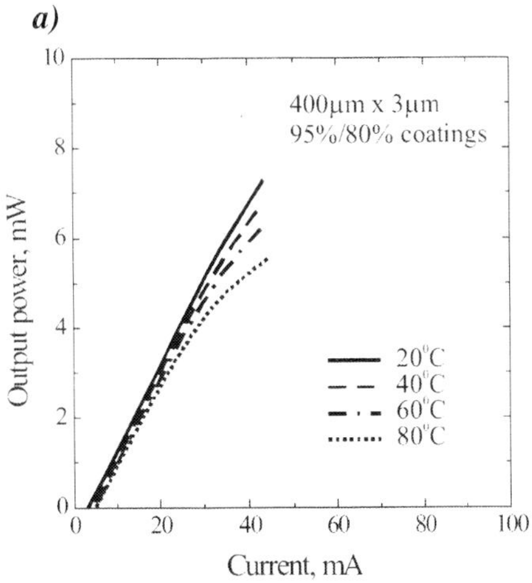

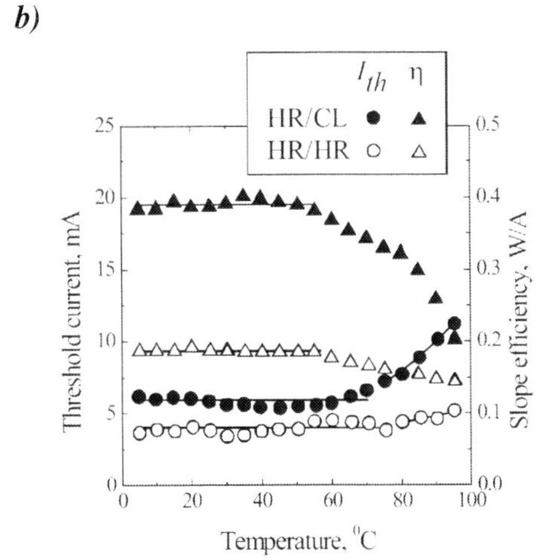

Figure 26. Light-current characteristics for 400-μm-long single mode 1.3-μm p-doped QD lasers at various temperatures (a) and variation of threshold current, I_{th} (circles), and slope efficiency, η (triangles), with temperature with HR/CL (solid symbols) or HR/HR (open symbols) facet coatings (b).

Using similar technique the p-doping of quantum dots in the active region have been carefully optimized by measuring the temperature dependence of the output (light-current) characteristics of QD-in-QW lasers [64]. The doping concentration was varied to provide sheet acceptor concentration per quantum dot layer varying in the range $(0\text{-}2)\times10^{12}$ cm^{-2}. The optimum doping level was determined by studying the luminescence of the dots and the device characteristics to be 5×10^{11} cm^{-2} (approximately 10 holes per single QD).

The light-current characteristics of a 3μm-wide 400μm-long single-mode laser with HR/HR facet coatings measured at various temperatures are shown in Figure 26,a. Plotted in Figure 26,b are the threshold current and differential efficiency as a function of temperature, as derived from the data of Figure 26,a. Data for the laser with one as-cleaved facet (HR/CL) are shown as well. It is evident that the threshold current is independent of temperature up to 75°C (i.e., $T_0=\infty$) and so the slope efficiency also remains constant in this temperature range. Similar results were obtained for the broad area lasers. In contrast, the conventional undoped QD lasers exhibit T_0=137K in the same temperature range (5-75°C). This is the first time that a semiconductor laser has displayed these features, characteristic only of atomic lasers.

Since the threshold currents and external quantum efficiencies of a diode lasers are usually sensitive to operating temperature, it is necessary to adjust bias and modulation currents in order to maintain good high-speed modulation characteristics in a wide temperature range. Significant improvements in temperature stability of laser characteristics by QDs give a chance to solve the mentioned problem. It was reported by group from University of Tokyo and Fujitsu labs in cooperation with NL-Nanosemiconductor that the small signal modulation characteristics of 1.3-μm QD laser with P-doped active region are also insensitive to operating temperature, as well as the static characteristics [65]. The maximum modulation bandwidth estimated at -3dB response of 8.1 and 7.7 GHz was obtained at 20 and 70°C, respectively. The relaxation oscillation frequency, f_r, is summarized in Figure 27 against the square root of the bias current above the threshold at different temperatures. Constant f_r is observed from 20 to 70°C in the range of low bias currents.

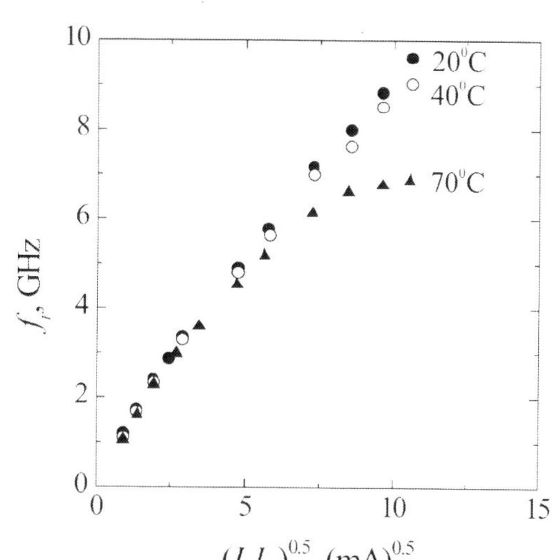

Figure 27. The bias-current dependence of the relaxation oscillation frequency at various temperatures for 1.3-μm QD laser.

Preliminary experiments on high-speed data transmission were fulfilled [66]. A 5-Gb/s data modulation and transmission was investigated using Fabry–Perot QD lasers emitting at 1.3 μm. Error-free transmission of 5-Gb/s data at room temperature over 4 km of single-mode fiber (SMF) and over 500 m of installed grade multimode fiber were demonstrated for the first time. Successful 10 Gb/s direct modulation from 20 °C to 70 °C without current adjustment have been reported very recently [65]. Clear eye opening with extinction ratio of 7dB was observed all over this temperature interval. The temperature-induced deviation of the averaged output power is as low as -0.2dB at 70°C while it was -1.5dB for InGaAsP QW laser, see Figure 28. The observed performance of QD lasers, far superior to QW lasers, is very promising for low-cost transmitters for 10 Gb/s Ethernet and OC-192/STM-64 applications.

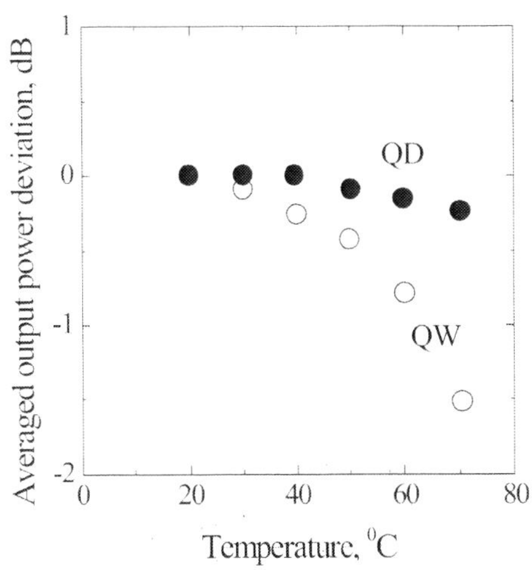

Figure 28. Deviation of averaged output power from the averaged output power at 20°C under 10 Gb/s modulation against operating temperature for 1.3-μm QD and QW lasers.

Several interesting features were found for QD active region that make it very promising for laser applications. One of such findings is a behavior of QD laser after high-temperature treatment. Annealing results in blue shift of the wavelength due to partial intermixing of In and Ga atoms from QDs and the matrix material [67]. Such a behavior was known before for QW lasers. However, in QD case the blue shift is very strong while the threshold current still remains rather low [68].

Figure 29,a shows room temperature lasing spectra of QD lasers made from different pieces of the same epiwafer subjected to different high-temperature treatment. Figure 29,b summarizes an evolution of lasing wavelength and threshold current density with duration of 700^{0}C annealing. In the initial (as-grown) structure the lasing wavelength was 1290 nm with the threshold current density about 250 A/cm^2. It is seen that initial 1.3-μm wavelength shifts to 0.9 μm after 4 hours of annealing. Thus, the interval of possible tuning is about 400 nm. The shortest wavelength of about 0.84 μm was achieved by 750^{0}C-annealing, Figure 29,a. The threshold current density is practically unchanged with the annealing as shown in Figure 29,b. Diode electrical properties, such as voltage of opening and series resistance, also remain undisturbed upon annealing.

The fact that significant blue shift caused by annealing does not accompany by degradation of threshold characteristics indicates high stability of QDs array to nucleation of structural defects. Such a behavior open a possibility for fabricating laser of different spectral ranges from single epiwafer just applying appropriate regimes of temperature treatment.

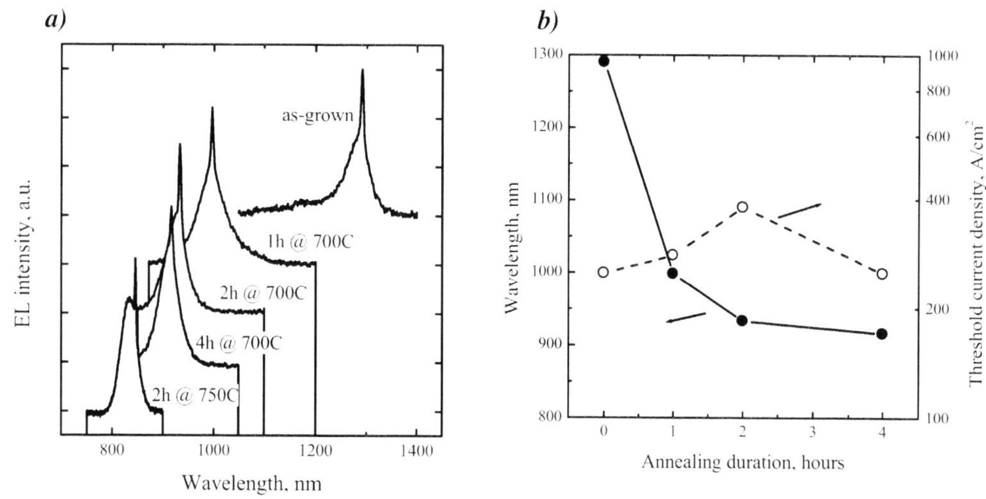

Figure 29. a) Room temperature lasing spectra; b) wavelength (solid circles) and threshold current density (open circles) against annealing duration for QD lasers made of the same epiwafer and annealed with different regimes.

Another interesting phenomenon is associated with suppression of in-plane carrier diffusion in the active region of QD laser [69]. Figure 30 demonstrates cross-section SEM images of single-mode QD lasers of different design. The same QD laser wafer was used. The only difference is the depth of etching. In one case the etching was stopped at about 0.4 μm above the waveguide (Figure 30,a), in the other case the etching profile entered the $Al_{0.13}Ga_{0.87}As$ waveguide region (Figure 30,b). In ultimate case of deep-mesa structure processed using a dry etching technique the process was stopped after 0.4 μm of the bottom $Al_{0.7}Ga_{0.3}As$ cladding layer was etched-off (Figure 30,c).

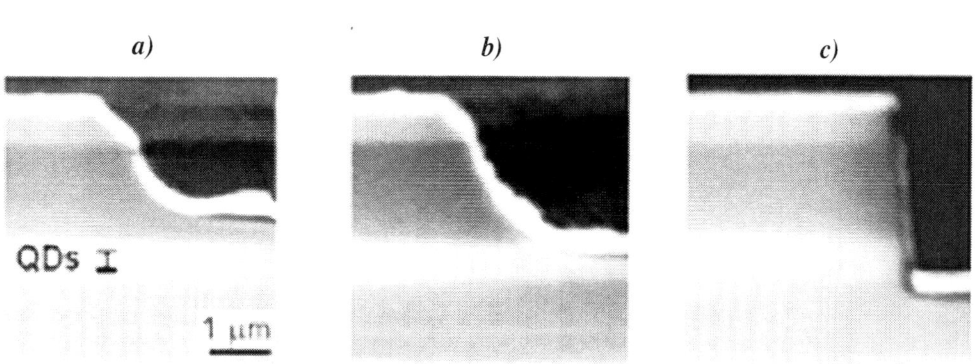

Figure 30. Cross-section SEM images of QD mesa structures fabricated by etching of different depths.

Light-current characteristics for these devices are shown in Figure 31. Maximum external differential efficiency of 50% and the lowest threshold current density (~125 A/cm^2) was observed for a deep-mesa laser. These observations were attributed to complete prevention of the current spreading. In opposite, some degradation of device characteristics is typically observed in a QW laser due to fast recombination on open sidewalls. For the QD laser, in-plane carrier diffusion is impeded by deep potential of quantum dots. In its turn, this results in effective suppression of sidewall nonradiative recombination. This finding is very important and opens a new way for cost-efficient fabrication of distributed feedback and photonic crystal QD devices.

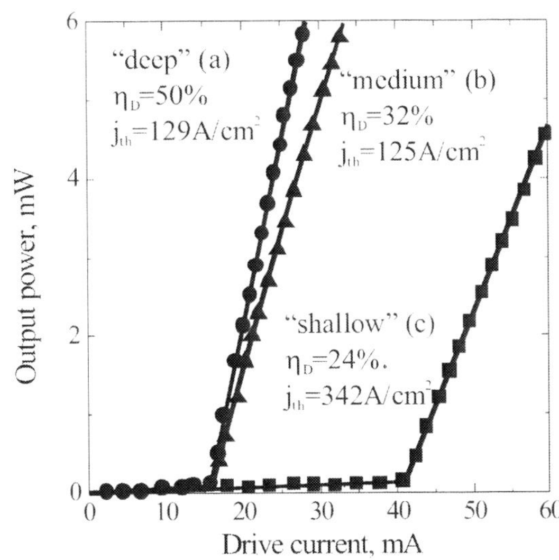

Figure 31. Room-temperature light-current characteristics for the 1.5mm-long 8μm-wide CL/HR devices fabricated with different etching profiles (see Figure 30).

In the rest of this section we would like to address issues of surface-emitting lasers based on self-organized QDs. Vertical-cavity surface-emitting laser (VCSEL) has several advantages over standard edge-emitters. They are simpler post-growth processing with a possibility of on-wafer test as well as possibility to form two-dimensional laser arrays. Also, epitaxial wafer is used more effectively in case of VCSEL processing: This means that more number of lasers can be fabricated from the same area. Intrinsically, VCSEL structure has better symmetry of the output beam and does not require sophisticated grating for wavelength stabilization. For a long time realization of long-wavelength QD VCSEL was difficult because of higher optical gain that is typically required for VCSEL active region. Development of multiply stacked QD arrays with sufficient saturated gain on QD ground state allowed demonstrating QD VCSELs with promising characteristics.

Careful optimization of the QD VCSELs in terms of minimizing the internal cavity losses and reduction of mirror transmission is critical point to obtain lasing in surface-emitting structure, especially for long-wavelength QD VCSELs [70]. In conventional 850 nm VCSELs p-doped and n-doped AlGaAs DBRs are usually exploited. This design provides easy device fabrication process, which is completely planar except one non-precise mesa etching (in the case of oxide-confined aperture). However, several studies showed that one of the main cavity loss mechanisms is related to free carrier absorption inside of DBRs, especially due to p-doping. These absorption losses limit the maximum possible reflectivity [71]. The possible alternative consists in using the intracavity contact design with high-reflectivity low-loss undoped AlGaAs/GaAs [72] or selectively oxidized Al_xO_y/GaAs DBRs [73].

First generation of QD VCSELs was based on intracavity contact design with selectively oxidized top and bottom DBRs and current apertures [74]. In these devices MBE-grown 3-layer stacked InAs/InGaAs QD structure was used as an active region. The 5□ cavity with two aperture layers is sandwiched between top (5 pairs) and bottom (7 pairs) AlGaO/GaAs DBRs. Two selectively oxidized current apertures provide not only current confinement but also additional redistribution of optical field inside the microcavity. Low internal optical loss for such devices resulted in relatively high value of external differential efficiency (>40%) even with high-reflection top oxidized DBR and simultaneously provided low threshold current density (1.2 mA for QD VCSEL with the aperture size of 12x12 μm^2) [75]. However, complicated fabrication process for such QD VCSELs and their high serial resistance (> 200Ω) due to necessity in low doping of the inracavity contact layers may be serious obstacle on the way to uncooled high-speed devices with reasonable output power (>1 mW).

Alternative design of QD VCSEL based on AlGaAs/GaAs all-semiconductor DBRs results in much simpler fabrication process [76]. Intracavity contacted VCSEL structure includes two selectively oxidized current apertures and AlGaAs/GaAs DBRs as shown schematically in Figure 32. Totally nine QD planes divided into three groups are located near the maxima of the optical field. The sets of QD layers were placed within the 2□-thick undoped GaAs layer and surrounded by n- and p-doped $Al_{0.98}Ga_{0.02}As$ quarter-wave aperture layers which were later selectively oxidized to form both current and waveguide apertures. These aperture layers are followed by 1.75□-thick intracavity contact/current spreading layers which are followed by 29 pairs top and 35 pairs bottom undoped $Al_{0.9}Ga_{0.1}As$/GaAs DBRs. The 1.75□-thick intra-cavity contact layer is p-doped with Be to $1x10^{18}$ cm^{-3} and includes two λ/16-thick Be spikes doped to $1x10^{19}$ cm^{-3} centered at the two standing wave nodes closest to the top DBR. Similarly, the lower n-doped 1.75□-thick intracavity contact layer is doped with Si to $1.5x10^{18}$ cm^{-3} and includes two λ/16-thick Si spikes doped to $4x10^{18}$ cm^{-3} centered at the two standing wave nodes closest to the bottom DBR.

The CW room temperature output light power-current-voltage (L-I-V) characteristic for fabricated QD VCSEL with 12 μm-diameter current aperture is shown in Figure 33 (solid lines). The threshold current of 2.2 mA and maximal output power as high as 2 mW at drive current of 4 mA were observed. Multimode output power is higher than 2 mW with slope efficiency of 0.41 W/A. For comparison the CW room temperature light power-current and current-voltage characteristics of a typical InAs-InGaAs QD VCSEL with AlO/GaAs DBRs are also shown (dotted lines). The structure with AlGaO/GaAs DBRs provides relatively short effective cavity length. In combination with relatively low uniform doping of intracavity contact layers (~ 10^{18} cm^{-3}) it should result in low internal optical loss. Additionally transparency current for 3-layer QD laser structures is sufficiently low as compared to 9-layer QD lasers. Above-mentioned factors resulted in smaller threshold current for QD VCSEL with oxidized DBRs (~ 1.2 mA). However, the maximal CW output power was less than 1mW even at room temperature. This is probably due to high serial resistance and low heat conductivity for oxidized DBRs.

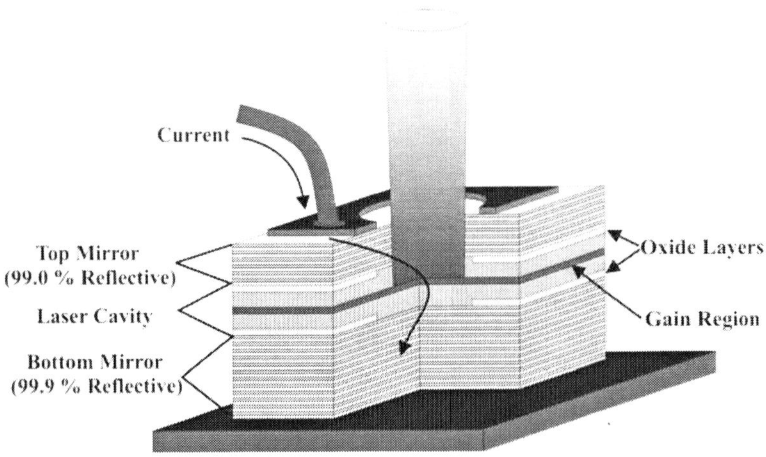

Figure 32. Schematic image of intracavity-contacted QD VCSEL on a GaAs substrate with oxidized current apertures.

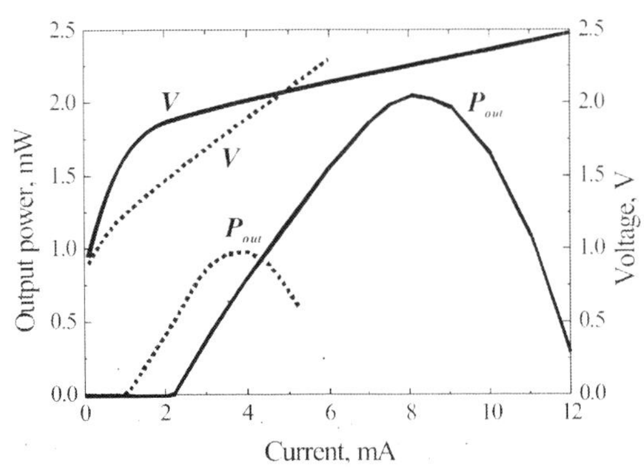

Figure 33. Room temperature output power, P_{out}, and voltage, V, against drive current for CW-operating intracavity-contacted QD VCSELs with oxidized (dotted lines) and semiconductor (solid lines) DBRs.

Simulation demonstrates that VCSEL structure with 9 QD layers provides about two times higher modal optical gain despite of relatively large effective cavity length. Moreover bottom semiconductor DBR should provide better heat dissipation than thr oxidized one. Combined with much lower serial resistance these factors resulted in about two times high value of maximal CW output power (~ 2mW). At the same time threshold current for QD VCSEL with semiconductor DBRs is about two times high (~2.2mA) but still low enough for practical applications. External differential quantum efficiency (43-46%) and maximum total wallplug efficiency (11-14%) were same for both VCSELs respectively and confirm possibility to keep low internal optical loss by using the optimized doping profile.

Further improvements in VCSEL scaling characteristics are expected for already existing QD structure. From this point of view non-uniformity in the sizes of individual QDs does not affect predicted improvements. As it was shown recently, both carrier transport in the lateral direction and carrier surface recombination are suppressed in QD structures [77].

4. Metamorphic Quantum Dot Structures

The longest lasing wavelength of 1.38 μm is reported for photopumped MOCVD InGaAs-GaAs QD structures [78]. More recently, MBE QD lasers at 1.35 μm with low threshold current density of 140A/cm^2 were realized [53]. Nevertheless, the wavelength of QD structures pseudomorphically grown on GaAs substrate is typically not longer than 1.3-1.32 μm. It would be highly desired to extend the spectral range of GaAs-based diode lasers to longer wavelength to meet the requirements of optical-fiber communication at the wavelengths near 1.55 μm.

Figure 34 shows the experimental dependence of the wavelength of QD emission on the bandgap of surrounding matrix. Different matrix bandgap supposes different materials. Three possible materials on GaAs substrate were previously known: pure bulky GaAs, bulky AlGaAs [31], and strained InGaAs QW. Also QDs in InGaAs matrix lattice-matched to InP substrates were studied [79]. For the given matrix material QD wavelength can be changed in some range by changing the effective thickness of InAs deposited. At the same time, changing the matrix bandgap provides much wide range of tunability: from 1 μm in case of AlGaAs up to 2μm for QDs on InP.

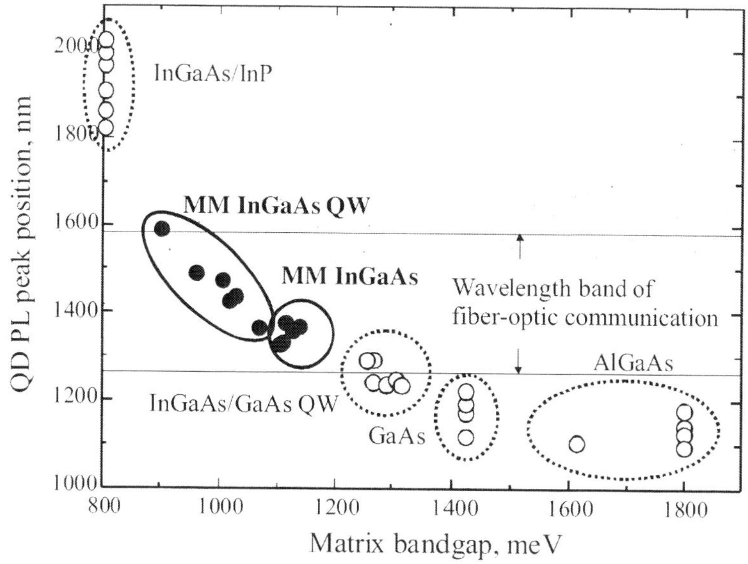

Figure 34. Wavelength of emission for InAs QDs embedded in different surrounding matrixes. Metamorphic matrices (MM, shown by solid circles) are discussed in Sec. 4. Two horizontal lines mark the spectral interval suitable for fiber-optic communication.

However there is a gap in possible matrix bandgap, near 1.0-1.1 eV, because no suitable material is available. This wavelength interval is the most interesting for optical fiber communication. Mentioned QDs in pseudomorphic InGaAs QW (QD-in-QW) only touch this region near 1.3-1.35 μm. Further increase of In mole fraction in the QW results in plastic strain relaxation. In this Section we consider a possibility to achieve sufficiently narrow-bandgap matrix for QD formation using the concept of metamorphic growth [80].

A possibility to exploit metamorphic growth mode for the long-wavelength lasers has not been studied in detail yet in spite of the fact that such an approach is currently widely used for high-electron mobility transistor (HEMT) structures (see, e.g. [81]). A virtual substrate, which has larger lattice parameters and lower bandgap energy as compared to the GaAs, can be created by the deposition of a sufficiently thick transient InGaAs buffer on an initial GaAs substrate. Purpose of the buffer is to change the lattice parameter from GaAs to InGaAs and simultaneously block all the dislocations inside. In ideal case the strain relaxation proceeds by misfit dislocations which propagate along the interface. The subsequent layers of the structure can be dislocation-free and, thus, be suitable for light-emitting devices.

Metamorphic approach allows one to extend the spectral range of QD structures grown on GaAs substrates using the sensitivity of QD wavelength to the bandgap of surrounding material. Figure 35 shows the compositional dependence of the bandgap for $In_x(Al_yGa_{1-y})_{1-x}As$ quaternary alloy assuming unstrained case. For In mole fraction, x, about 20-25% the badngap interval from 1.1 to 1.8 eV can be covered by changing the Al subfraction, y. This wide range provides both sufficiently narrow-bandgap matrix for QD formation (InGaAs) and effective prevention of current leakage when using InGaAlAs cladding layers (y>0.3). A comparison of the band diagrams of the 1.3-μm (dashed line) and the 1.5-μm (solid line) QD lasers is shown in Figure 36. Layer sequence of the metamorphic-based 1.5-μm laser looks very similar to the conventional 1.3-μm QD laser on GaAs with adding approximately 20-25% of InAs to all the layers. The structure is initialized from a metamorphic InGaAs buffer layer.

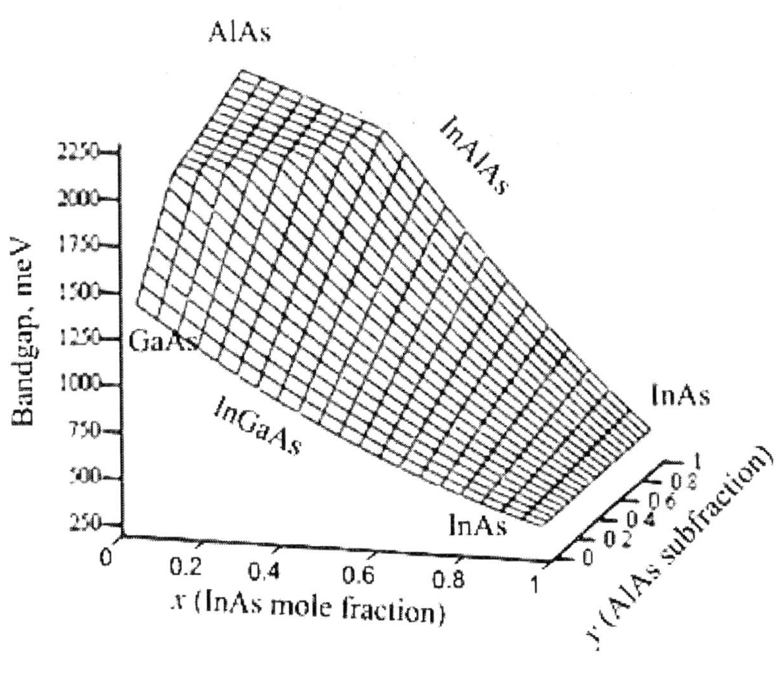

Figure 35. Bandgap of $In_x(Al_yGa_{1-y})_{1-x}As$ quaternary alloy (300K) as a function of InAs and AlAs fractions.

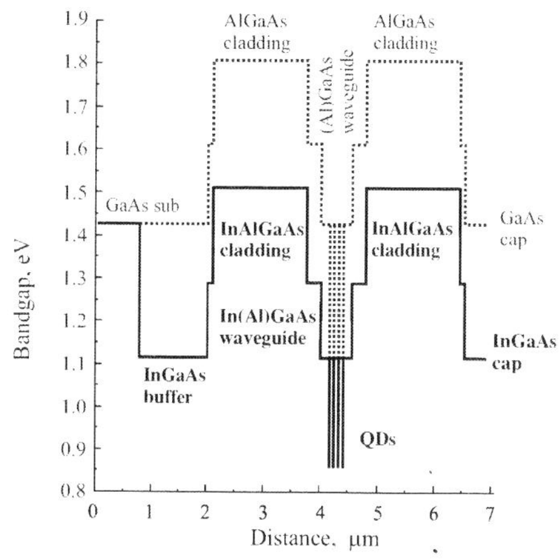

Figure 36. Bandgap profile of 1.3-μm (dashed line) and 1.5-μm (solid line) QD laser structures. In the latter case the pseudomorphic growth mode is supposed.

Structural quality, degree of strain relaxation, surface morphology and dislocation density of metamorphic epilayers were evaluated by combination of high-resolution X-ray diffraction, scanning and transmission electron microscopy methods [82]. The discussed structures were grown by molecular-beam epitaxy (MBE) on GaAs (100) substrates. $In_xGa_{1-x}As$ epilayers of different In mole fraction, x, ranging from 16 to 26% of 2-μm thick were deposited onto GaAs at different growth conditions which include linear or step compositional gradient, various temperature of deposition, As overpressure, etc. Si- and Be-doped InAlGaAs epilayers were studied as well to ensure suitable electrical characteristics.

In opposite to ideal metamorphic growth, formation of both interface and threading dislocations is typically observed in the transient buffer layer. Threading dislocations are especially dangerous from the viewpoint of possible light-emitting applications because of their propagation through the whole thickness of the epitaxial layer. It was found that interface-to-threading dislocations ratio depends on the growth conditions. Formation of the threading dislocations can be effectively eliminated by careful optimization of the structure design and growth regimes. As an example, TEM images of the samples grown at different temperatures are presented in Figure 37. It is seen that that decrease of epitaxial temperature results in domination of interface-type dislocations while the threading those are suppressed. Use of lowered temperatures for the layer deposition ($\leq 400^0$C) allows one to confine most of dislocations inside the buffer and avoid their propagation into the upper layers.

Surface morphology depends to a certain extent on temperature of buffer deposition. The growth mode switches to the island type if too low temperatures ($\leq 350^0$C) are used which is also confirmed by spotty RHEED pattern. Two-step temperature regime, when the initial buffer is deposited at 400^0C and the rest of structure at 500^0C, provides smooth layer surface in combination with filtering of threading dislocations. The dislocation density in the upper layers of about 2×10^8 cm^{-2} was estimated by plan-view TEM images. The degree of strain relaxation is as high as 99.7% as evaluated by HRXRD curves taken near asymmetric (115) GaAs reflexes.

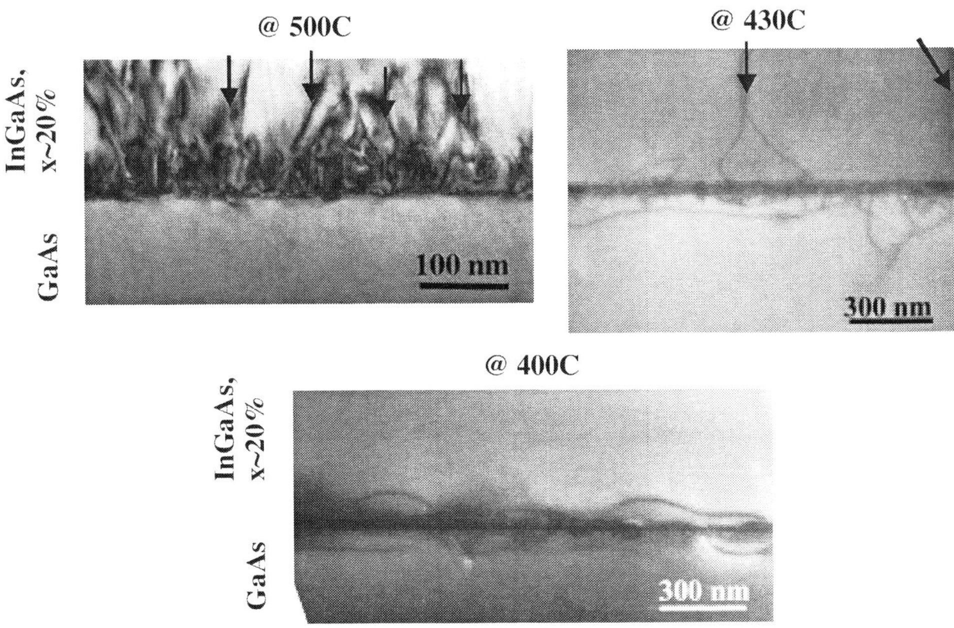

Figure 37. Cross-section TEM images of $In_{0.2}Ga_{0.8}As$ metamorphic layers deposited on GaAs surface at different temperatures. Some threading dislocations are marked by black arrows.

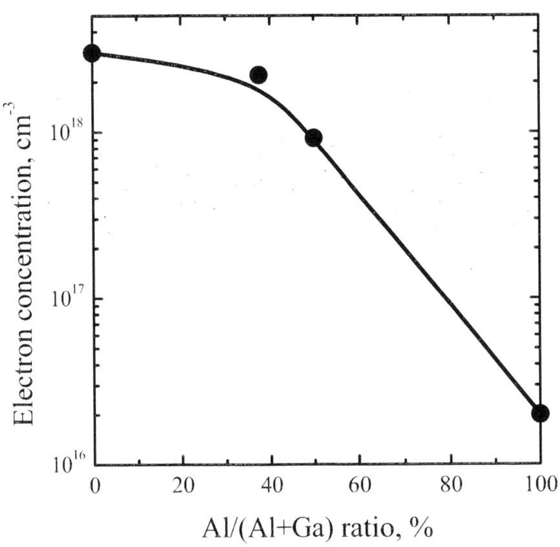

Figure 38. Dependence of 300K electron concentration in Si-doped $In_{0.2}(Al_yGa_{1-y})_{0.8}As$ metamorphic epilayers on Al subfraction.

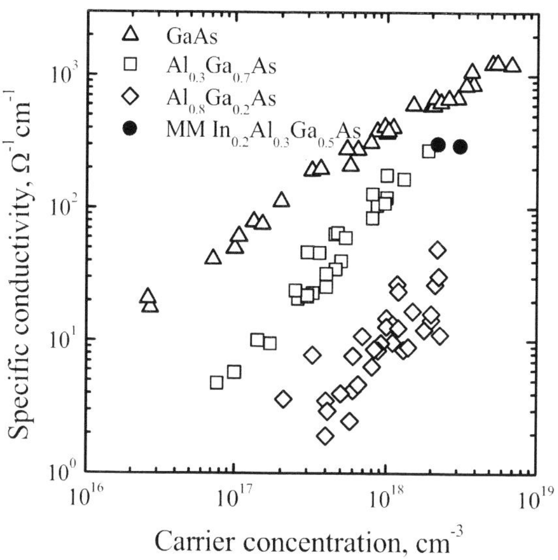

Figure 39. Correlation between carrier concentration and specific conductivity in Si-doped GaAs (triangles), $Al_{0.3}Ga_{0.7}As$ (squares), $Al_{0.8}Ga_{0.2}As$ (rhombs), and metamorphic $In_{0.2}Al_{0.3}Ga_{0.5}As$ (solid circles).

Conductivity of Be- and Si-doped $In_x(Al_yGa_{1-y})_{1-x}As$ metamorphic layers (x~20-25%) was studied to make a conclusion concerning applicability of metamorphic layers for fabricating components of laser structure. It was shown that p-type doped layers demonstrate free-hole concentration in a good agreement with that of GaAs:Be regardless the Al composition. At the same time, free-electron concentration in n-type doped epilayer with x~20% decreases with increasing Al subfraction, y. Room-temperature concentration against Al subfraction, y, is shown in Figure 38. A significant problem with conductivity was found for the samples with y>50% (two order of magnitude drop in the electron concentration as compared to the concentration of Si atoms).

On the other hand, the samples with lower Al content ($y \leq 40\%$) can be effectively doped in excess of 10^{18} cm^{-3} while the electron mobility is about 800-900 cm^{2}V^{-1}s^{-1}. As a result, the specific conductivity of $In_{0.2}(Al_yGa_{1-y})_{0.8}As$:Si ($y$~30%) metamorphic cladding layers can reach 300 Ω^{-1}cm^{-1}. This is much higher than that of $Al_{0.8}Ga_{0.2}As$ and corresponds to the highest conductivity level of $Al_{0.3}Ga_{0.7}As$:Si (see Figure 39), two the most common materials for cladding layers in 1.3-μm QD lasers.

Formation of island array on top of metamorphic InGaAs layer was confirmed by transition of RHEED pattern from streaky to spotty upon deposition of approximately 1.8 ML of InAs [80]. TEM micrograph of InAs QD array formed in a metamorphic $In_{0.2}Ga_{0.8}As$ matrix is presented in Figure 40. Quantum dots are approximately 15-20 nm in diameter with the surface density of about 10^{11} cm^{-2}.

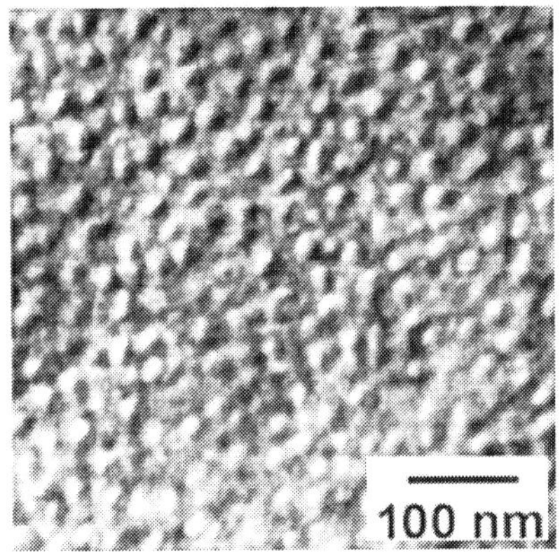

Figure 40. Plan-view TEM image of InAs QD array formed in a metamorphic In0.2Ga0.8As matrix.

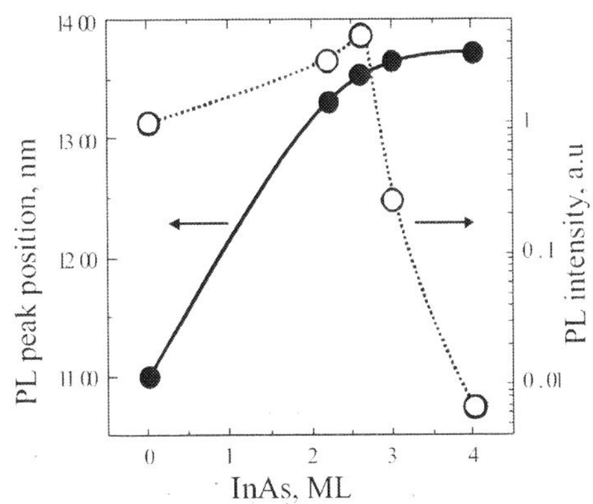

Figure 41. Dependence of room-temperature PL peak position (solid symbols) and PL intensity (open symbols) on the effective thickness of InAs forming QD array in a metamorphic $In_{0.2}Ga_{0.8}As$ matrix.

Dependence of room-temperature PL peak position and PL intensity on the effective thickness of InAs forming QD array is shown in Figure 41. The bandgap of a metamorphic matrix was about 1.1 eV. In the QD growth mode (2-4 ML) PL peak position is slightly red-shifted with increasing the InAs thickness from 1.33 to 1.37μm indicating the growth of QDs in size. Wavelength of QDs in a metamorphic matrix is approximately 150 nm longer as compared to that of InAs/GaAs QDs for the same amount of InAs (compare with data of Figure 5,b). This is a direct consequence of smaller bandgap energy of the matrix material. Intensity of QD line initially increases (up to ~2.7ML of InAs) and then abruptly drops due to plastic strain relaxation.

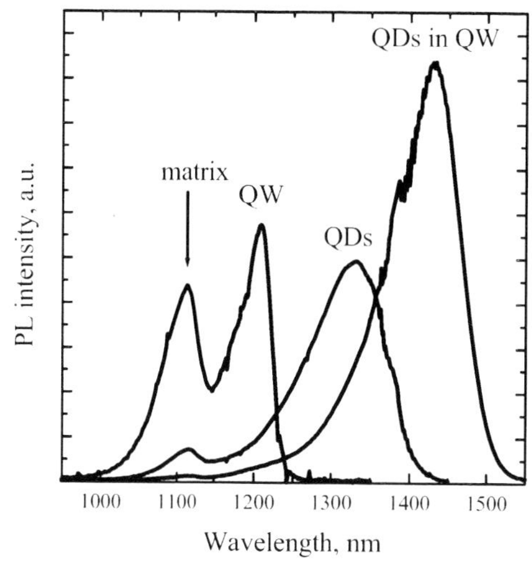

Figure 42. Room-temperature PL spectra from 4-nm-thick $In_{0.4}Ga_{0.6}As$ quantum well (QW), 2.7ML InAs quantum-dot array (QDs), and 2.7ML QD array capped with $In_{0.4}Ga_{0.6}As$ QW (QDs in QW) having the same matrix bandgap of 1.12 eV.

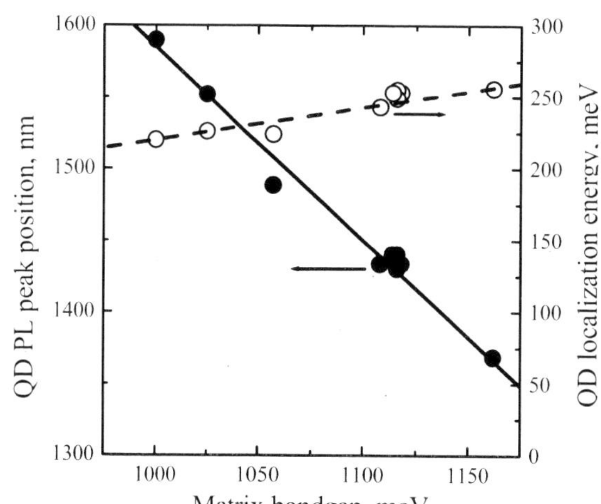

Figure 43. Dependence of QD emission wavelength (solid circles) and QD localization energy (open circles) on the bandgap of metamorphic $In_xGa_{1-x}As$ matrix. QD array is capped with 4-nm-thick $In_{x+0.2}Ga_{0.8-x}As$ QW.

Wavelength of QD emission can be shifted further (beyond 1.4 μm) by capping InAs QD array with InGaAs QW. Similar method was discussed in Sec. 2 for achieving the 1.3-μm emission in a GaAs matrix (QD-in-QW). In the discussed case of metamorphic structures InAs composition in the capping QW is about 40%, i.e. 20% higher than that in the metamorphic matrix; QW width is 4 nm. PL spectra of QW, QD array, and QD array capped with QW are compared in Figure 42 for the matrix bandgap of 1110 meV. For the same matrix bandgap this QD-in-QW approach results in additional 130nm shift of QD line to longer wavelengths.

Dependence of QD emission wavelength on the bandgap of metamorphic $In_xGa_{1-x}As$ matrix is shown in Figure 43. QD line can be controllably redshifted up to 1.6μm by varying the matrix composition (matrix bandgap). Thus, QD structures grown in a metamorphic matrix completely cover the wavelength interval of optical fiber communication (see Figure 34). With increasing InAs mole fraction in the matrix (decreasing the matrix bandgap) the energy separation between QD line and the matrix bandgap slightly decreases (Figure 43). However, the effective localization energy exceeds 200 meV even for the longest wavelength-emitting QDs. This should be sufficient for effective prevention of thermally activated carrier escape out of QD states to the matrix at room temperature.

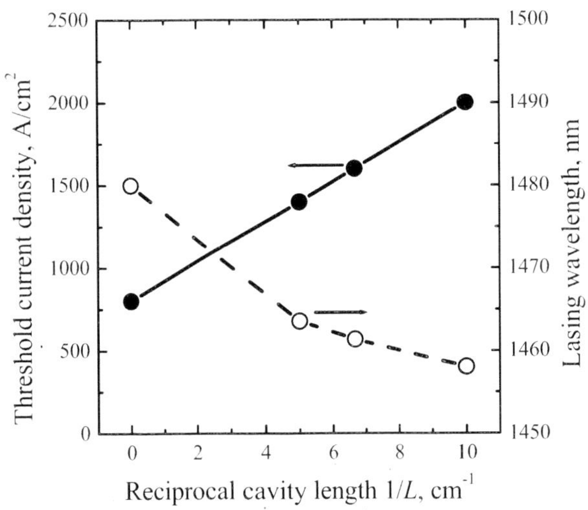

Figure 44. Threshold current density (solid circles) and lasing wavelength (open circles) against reciprocal length of broad-area laser with 10 planes of metamorphic QDs.

The metamorphic approach was applied for fabricating long-wavelength lasers on GaAs substrates [83]. InAs mole fraction throughout the structure was 21% excluding a QD active region. 1.6-μm-thick InAlGaAs claddings doped with Si or Be had Al composition of 35%. A 0.7-μm-thick undoped InGaAs layer was used as a waveguide as well as a matrix for QD formation. From 5 to 15 QD planes were deposited in center of the waveguide with separation by 45-nm-thick spacer layers. Each QD layer was formed by deposition of 2.7 ML of InAs and covered with 4-nm-thick $In_{0.41}Ga_{0.59}As$ QW. The structure was terminated by Be-doped InGaAs contact layer. Broad-area lasers with a 100 μm stripe width were fabricated. No facet coatings were deposited. N- and p-type contacts were made using deposition and melting at 450°C of AuGe/Ni/Au and AuZn/Ni/Au metallic layers, respectively. Laser characteristics were measured in the 20-85°C temperature interval under 0.2 s pulse excitation. To evaluate internal device parameters the mirror loss was varied by changing the cavity length, L. Also, laser structures with four cleaved facets were evaluated. Such kind of laser cavity emulates the stripe laser of infinite length ($L^{-1} \approx 0$) because the mirror loss is negligible.

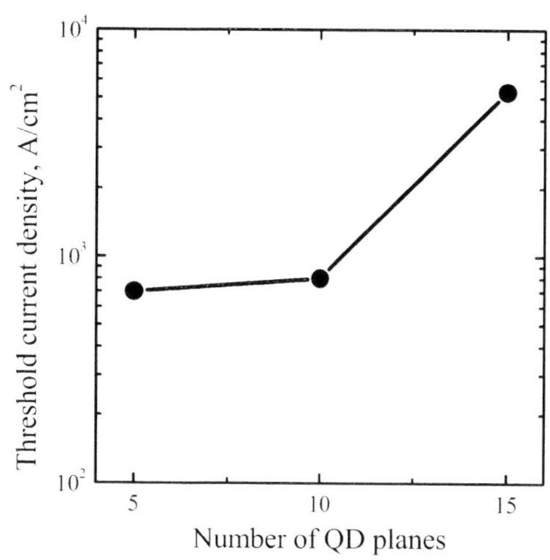

Figure 45. Dependence of the minimum threshold current density (resonators with four cleaved facets) in the number of QD planes in the laser active region.

Threshold current density measured at room temperature and lasing wavelength taken near the threshold are shown in Figure 44 as functions of the reciprocal cavity length for the structure with 10 QD planes. Minimum threshold current density of 800 A/cm^2 and the longest wavelength of 1.49 μm are achieved in the resonator with four cleaved facets. Lasing proceeds on the ground state up to 1-mm-long cavity. This means that the saturated gain on the ground state exceeds 10 cm^{-1}. With varying the cavity length from 2 to 1 mm the threshold current density increases from 1.4 to 2 kA/cm^2 while the wavelength changes very slightly (from 1463 to 1458 nm). In all lasers the lasing peak is shifted towards the longer wavelength with respect to PL peak position at 1.45 μm measured in a sample with etched-off contact layer. This also indicates a sufficient gain on the QD ground state.

Number of QD planes in the laser active region was varied to study its impact on threshold characteristics. Dependence of the minimum threshold current density (resonators with four cleaved facets) on the number of QD planes is shown in Figure 45. The lowest threshold current density of 700 A/cm^2 was measured in the laser structure with 5 planes of metamorphic QDs as compared to 800 A/cm^2 in the 10-plane laser structure. In case of 15 QD planes the threshold current density increases to 5.3 kA/cm^2. It was attributed to formation of dislocations in the active region due to excessive strain.

Example of light-current characteristic of a representative metamorphic QD laser is shown in Figure 46. The maximum power in pulse regime exceeds 7W. The lasing spectra shown in the insert demonstrate the absence of any current-induced shift of the wavelength. Up to the maximum drive current of 23A quantum dots lase on the ground state with the wavelength of 1.46 μm. The external differential efficiency is 50%. From the dependence of the external differential efficiency on cavity length the following internal parameters were extracted: internal efficiency as high as 60% and internal loss of ~3 cm^{-1}.

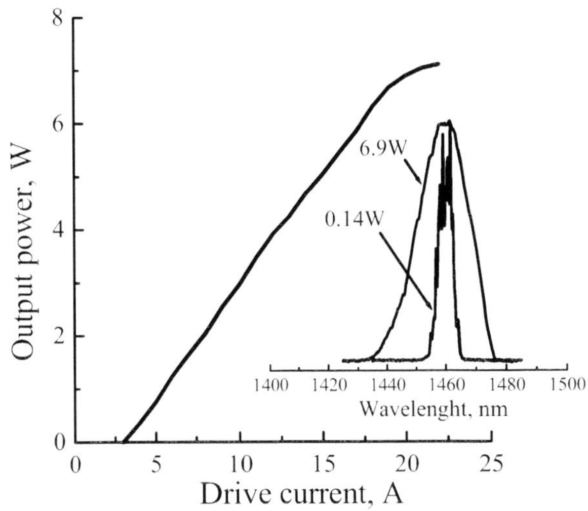

Figure 46. Light-current characteristic in pulse regime and lasing spectra (insert) for broad-area metamorphic QD laser.

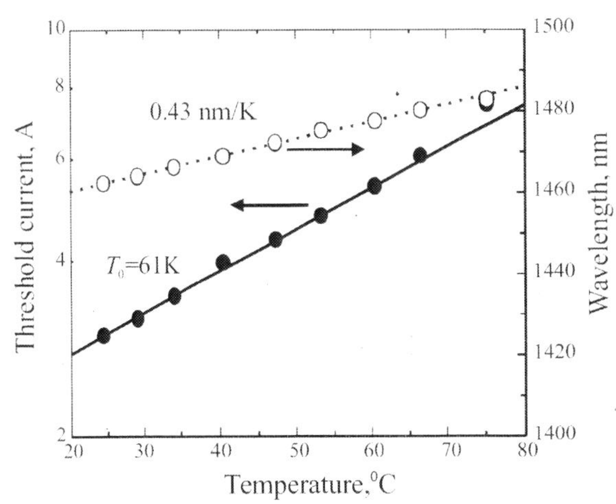

Figure 47. Temperature dependence of the threshold current (solid circles) and lasing wavelength (open circles).

Temperature dependence of the threshold current and lasing wavelength is presented in Figure 47. Maximum operation temperature comes to 80^0C, the characteristic temperature of the laser threshold, T_0, is 61K. Lasing wavelength is gradually red-shifted with increasing temperature. No temperature-induced transition to the excited-state lasing is observed. The lasing wavelength reaches 1515 nm in the diode with four cleaved facets and 1485 nm in case of 1.5-mm-long stripe laser.

Issues of design and growth of surface-emitting lasers on metamorphic QDs were preliminary studied [84]. VCSEL structure basically comprises bottom-side and top-side distributed Bragg reflectors (DBRs) which confine the laser microcavity with the active region inside. The active region of sufficiently long-wavelength (λ>1.45 μm) VCSEL structure in the metamorphic approach represents several planes of self-organized QDs embedded into strain-relaxed InGaAlAs matrix grown on metamorphic InGaAlAs/GaAs transient layer. The main question of a metamorphic VCSEL is a proper design of DBRs in order to ensure their high structural and optical quality and minimize possible negative effect on VCSEL characteristics.

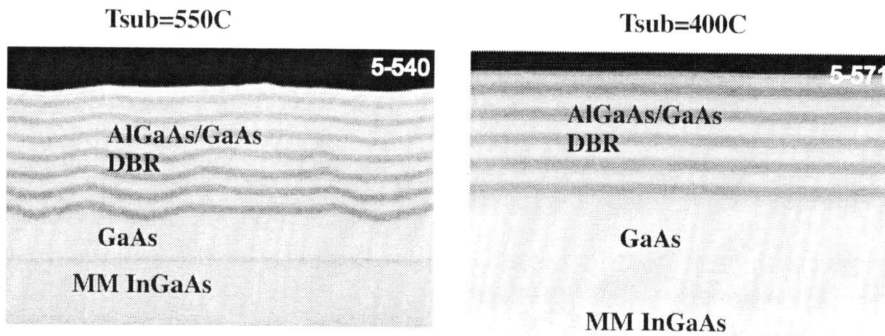

Figure 48. Cross-section TEM images of AlGaAs/GaAs DBR grown on strain-relaxed $In_{0.2}Ga_{0.8}As$ buffer at different temperatures.

If the metamorphic transient layer is located in a buffer, already bottom DBR should represent a sequence of metamorphic (strain-relaxed) pairs of InAlAs and InGaAs. However, optical contrast between $In_{0.2}Al_{0.8}As$ and $In_{0.2}Ga_{0.8}As$ is smaller as compared to conventionally used $Al_XGa_{1-X}As$/GaAs pairs (x~90-100%) and may be insufficient to achieve high reflectivity. Also, very thick metamorphic layer may lead to surface corrugation affecting formation of QDs and reflectivity of the DBR. In opposite, if the transient metamorphic layer is located inside the VCSEL cavity, conventional high-contrast AlGaAs/GaAs pairs can be used as bottom DBR. Also, formation of the QD active region proceeds under conditions which were previously tested in edge-emitting lasers.

Use of AlGaAs/GaAs pairs as a top DBR provides flexible VCSEL design suitable for both surface- and substrate-emitting devices with oxidized or oxide-free AlGaAs. Such a design can be realized by an inverse metamorphic transient layer inserted in top part of the cavity. This means that GaAs or AlGaAs is deposited on strain-relaxed InGa(Al)As. Preliminary study of metamorphic VCSEL structures demonstrated significant effect of growth temperature on planarity of AlGaAs/GaAs DBR grown by such a manner. Examples of cross-section electron-microscopy images are shown in Figure 48. If inverse metamorphic step is carried out at typical epitaxial temperatures of 500-600^{0}C very rough interfaces are formed (left-hand side image). At the same time, lowered deposition temperature of 400-450^{0}C ensures minimal corrugation (right-hand side image).

5. Conclusion

Since the beginning of 90s very fast progress is observed in a field of self-organized quantum dots. Optically active quantum-sized islands were synthesized in a number of material systems and studied in detail. Quantum dot diode lasers have been synthesized demonstrating on-going improvement of their device characteristics. Wide frontiers are expected for device application of QDs emitting in wavelength ranges beyond the spectral limit of quantum well. In particular, long-wavelength QD lasers grown on GaAs substrates offers unique combination of QD active region with extremely deep localization of the active region with respect to the matrix. Now, 1.3-μm QD diode lasers with the record-low threshold current densities were reported on. Also, very promising efficiency characteristics of QD lasers have been reported with the best external differential efficiencies in the 85-90% interval. Device performance of single-mode edge-emitting and surface-emitting (VCSEL) lasers meets the requirements of fiber optical communication technology. Further extension of the spectral range of GaAs-based quantum-dots lasers has been done by introduction of self-organized quantum dots in a metamorphic matrix. Metamorphic approach allows one to reach the 1.6μm interval at room temperature using the sensitivity of QD wavelength to the bandgap of surrounding material. Threshold current density and lasing wavelength of the metamorphic QD lasers corresponds fairly well to the best results published for long-wavelength lasers on GaAs based on InGaAsSbN [85] with the advantage of higher slope efficiency. These structures may have potential in edge- and surface-emitting lasers for applications in metropolitan area networks and in dense wavelength division multiplexing. Thus, QD laser technology have demonstrated its potential for the whole spectral range of optical fiber communication.

Acknowledgements

This work was supported by Russian Foundation of Basic Research, Programs of Ministry of Science of Russian Federation INTAS, EU Project "Sandie", and Joint Research Project between A. F. Ioffe Physico-Technical institute, St. Petersburg, Russia and NL-Nanosemiconductor-GmbH, Dortmund, Germany.

References

[1] Kondow, M.; Uomi, K.; Niwa, A.; Kitatani, T.; Watahiki, S.; Yazawa, Y. *Jpn. J. Appl. Phys.* 1996, 35, 1273-1275.

[2] Egorov, A. Yu.; Zhukov, A. E.; Kop'ev, P. S.; Ledentsov, N. N.; Maximov, M. V.; Ustinov, V. M.; Tsatsul'nikov, A. F.; Fedorov, D. L.; Bimberg, D.; Alferov, Zh. I. *Semiconductors*. 1996, 30, 707-710.

[3] Dingle R and Henry C H. Quantum effects in heterostructure lasers, *U.S. Patent* 3982207, (1976).

[4] Arakawa Y.; Sakaki, H. *Appl. Phys. Lett.* 1992, 40, 939–941.

[5] Goldstein, L.; Glas, F.; Marzin, J. Y.; Charasse, M. N.; Le Roux, G. *Appl. Phys. Lett.* 1985, 47, 1099-1101.

[6] Matthews, J. W.; Blakeslee, A. E. *J. Cryst. Growth.* 1974, 27, 118-125.

[7] Leonard, D.; Pond, K.; Petroff, P. M. *Phys. Rev. B.* 1995, 50, 11687-11692.

[8] Egorov, A. Yu.; Zhukov, A. E.; Kop'ev, P. S.; Ledentsov, N. N.; Maksimov, M. V.; Ustinov, V. M. *Semiconductors*. 1994, 28, 809-811.

[9] Ruvimov, S.; Werner, P.; Scheerschmidt, K.; Goesele, U.; Heydenreich, J.; Richter, U.; Ledentsov, N. N.; Grundmann, M.; Bimberg, D.; Ustinov, V. M.; Egorov, A. Yu.; Kop'ev, P. S.; Alferov, Zh. I. *Phys. Rev. B.* 1995, 51, 14766–14769.

[10] Lentzen, M.; Gerthsen, D.; Forster, A.; Urban, K. *Appl. Phys. Lett.* 1992, 60, 74-76.

[11] Leonard, D.; Krishnamurthy, M.; Reaves, L. M.; Den Baars, S. P.; Petroff, P. M. *Appl. Phys. Lett.* 1993, 63, 3203-3205.

[12] Moison, J. M.; Houzay, F.; Barthe, F.; Leprince, L.; Andre, E.; Vatel, O. *Appl. Phys. Lett.* 1994, 64, 196-198.

[13] Grundmann, M.; Christen, J.; Ledentsov, N. N.; Bohrer, J.; Bimberg, D.; Ruvimov, S. S.; Werner, P.; Richter, U.; Goesele, U.; Heydenreich, J.; Ustinov, V. M.; Egorov, A. Yu.; Zhukov, A. E.; Kop'ev, P. S.; Alferov, Zh. I. *Phys. Rev. Lett.* 1995, 74, 4043-4046.

[14] Kudo M.; Mishima, T. *J. Appl. Phys.* 1995, 78, 1685-1688.

[15] Ledentsov, N. N.; Shchukin, V. A.; Bimberg, D.; Ustinov, V. M.; Cherkashin, N. A.; Musikhin, Yu. G.; Volovik, B. V.; Cirlin, G. E.; Alferov, Zh. I. *Semicond. Sci. Technol.* 2001, 16, 502–506.

[16] Dubrovskii, V. G.; Musikhin, Yu. G.; Cirlin G. E.; Egorov, V. A.; Polyakov, N. K.; Samsonenko, Yu. B.; Tonkikh, A. A.; Kryzhanovskaya, N. V.; Bert, N. A.; Ustinov, V. M. *Semiconductors.* 2004, 38, 329-334.

[17] Mukai, K.; Ohtsuka, N.; Sugawara, M.; Yamazaki, S. *Jpn. J. Appl. Phys.* 1994, 33, L1710-L1712.

[18] Mirin, R. P.; Ibbetson, J. P.; Nishi, K.; Gossard, A. C.; Bowers, J. E. *Appl. Phys. Lett.* 1995, 67, 3795-3797.

[19] Huffaker, D. L.; Deppe, D. G. *Appl. Phys. Lett.* 1998, 73, 520-522.

[20] Zhukov, A. E.; Kovsh, A. R.; Ustinov, V. M.; Egorov, A. Yu.; Ledentsov, N. N.; Tsatsul'nikov, A. F.; Maximov, M. V.; Shernyakov, Yu. M.; Kopchatov, V. I.; Lunev, A. V.; Kop'ev, P. S.; Bimberg, D.; Alferov, Zh. I. *Semicond. Sci. Technol.* 1999, 14, 118–123.

[21] Zhukov, A. E.; Egorov, A. Yu.; Kovsh, A. R.; Ustinov, V. M.; Ledentsov, N. N.; Maximov, M. V.; Tsatsul'nikov, A. F.; Zaitsev, S. V.; Gordeev, N. Yu.; Kop'ev, P. S.; Bimberg, D.; Alferov, Zh. I. *Semiconductors.* 1997, 31, 411-414.

[22] Nishi, K.; Saito, H.; Sugou, S.; Lee, J.-S. *Appl. Phys. Lett.* 1999, 74, 1111-1113.

[23] Ustinov, V. M.; Maleev, N. A.; Zhukov, A. E.; Kovsh, A. R.; Egorov, A. Yu.; Lunev, A. V.; Volovik, B. V.; Krestnikov, I. L.; Musikhin, Yu. G.; Bert, N. A.; Kop'ev, P. S.; Alferov, Zh. I.; Ledentsov, N. N.; Bimberg, D. *Appl. Phys. Lett.* 1999, 74, 2815-2817.

[24] Lester L. F.; Stinz, A.; Li, H.; Newell, T. C.; Pease, E. A.; Fuchs, B. A.; Malloy, K. J. *IEEE Photon. Technol. Lett.* 1999, 11, 931-933.

[25] Saito, H.; Nishi, K.; Sugou, S. *Appl. Phys. Lett.* 1998, 73, 2742-2744.

[26] Volovik, B. V.; Tsatsul'nikov, A. F.; Bedarev, D. A.; Egorov, A. Yu.; Zhukov, A. E.; Kovsh, A. R.; Ledentsov, N. N.; Maximov, M. V.; Maleev, N. A.; Musikhin, Yu. G.; Suvorova, A. A.; Ustinov, V. M.; Kop'ev, P. S.; Alferov, Zh. I.; Bimberg, D. *Semiconductors.* 1999, 33, 901-905.

[27] Ustinov, V. M.; Zhukov, A. E.; Kovsh, A. R.; Maleev, N. A.; Mikhrin, S. S.; Tsatsul'nikov, A. F.; Maximov, M. V.; Volovik, B. V.; Bedarev, D. A.; Kop'ev, P. S.; Alferov, Zh. I.; Vorob'ev, L. E.; Firsov, D. A.; Suvorova, A. A.; Soshnikov, I. P.; Werner, P.; Ledentsov, N. N.; Bimberg, D. *Microelectronics J.* 2000, 31, 1-7.

[28] Ledentsov, N. N.; Ustinov, V. M.; Egorov, A. Yu.; Zhukov, A. E.; Maximov, M. V.; Tabatadze, I. G.; Kop'ev, P. S. *Semiconductors.* 1994, 28, 832-834.

[29] Kirstaedter, N.; Ledentsov, N. N.; Grundmann, M.; Bimberg, D.; Ustinov, V. M.; Ruvimov, S. S.; Maximov, M. V.; Kop'ev, P. S.; Alferov, Zh. I.; Richter, U.; Werner, P.; Gösele, U.; Heydenreich, J. *Electron. Lett.* 1994, 30, 1416-1417.

[30] Egorov, A. Yu.; Zhukov, A. E.; Kop'ev, P. S.; Ledentsov, N. N.; Maximov, M. V.; Ustinov, V. M.; Tsatsul'nikov, A. F.; Bert, N. A.; Kosogov, A. O.; Bimberg, D.; Alferov, Zh. I. *Semiconductors*. 1996, 30, 879-883.

[31] Zhukov, A. E.; Ustinov, V. M.; Egorov, A. Yu.; Kovsh, A. R.; Tsatsul'nikov, A. F.; Maximov, M. V.; Ledentsov, N. N.; Zaitsev, S. V.; Gordeev, N. Yu.; Kopchatov, V. I.; Shernyakov, Yu. M.; Kop'ev, P. S.; Bimberg, D.; Alferov, Zh. I. *J. Electron. Mat.* 1998, 27, 106-109.

[32] Schmidt, O. G.; Kirstaedter, N.; Ledentsov, N. N.; Mao, M. H.; Bimberg, D.; Ustinov, V. M.; Egorov, A. Y.; Zhukov, A. E.; Maximov, M. V.; Kop'ev, P. S.; Alferov, Zh. I. *Electron. Lett.* 1996, 32, 1302-1304.

[33] Ustinov, V. M.; Egorov, A. Yu.; Kovsh, A. R.; Zhukov, A. E.; Maksimov, M. V.; Tsatsul`nikov, A. F.; Gordeev, N. Yu.; Zaitsev, S. V.; Shernyakov, Yu. M.; Bert, N. A.; Kop`ev, P. S.; Alferov, Zh. I.; Ledentsov, N. N.; Bohrer, J.; Bimberg, D.; Kosogov, A. O.; Werner, P.; Gosele, U. *J. Cryst. Growth*. 1997, 175/176, 689-695.

[34] Zaitsev, S. V.; Gordeev, N. Yu.; Sherniakov, Yu. M.; Ustinov, V. M.; Zhukov, A. E.; Egorov, A. Yu.; Maximov, M. V.; Kop`ev, P. S.; Alferov, Zh. I.; Ledentsov, N. N.; Kirstaedter, N.; Bimberg, D. *Superlat. Microstruct.* 1997, 21, 559-564.

[35] Shernyakov, Yu. M.; Egorov, A. Yu.; Zhukov, A. E.; Zaitsev, S. V.; Kovsh, A. R.; Krestnikov, I. L.; Lunev, A. V.; Ledentsov, N. N.; Maximov, M. V.; Sakharov, A. V.; Ustinov, V. M.; Zhao, Zh.; Kop'ev, P. S.; Alferov, Zh. I.; Bimberg, D. *Tech. Phys. Lett.* 1997, 23, 149-150.

[36] Huffaker, D. L.; Park, G.; Zou, Z.; Shchekin, O. B.; Deppe, D. G. *Appl. Phys. Lett.* 1998, 73, 2564-2566.

[37] Park, G.; Shchekin, O. B.; Huffaker, D. L.; Deppe, D. G. *IEEE Photon. Technol. Lett.* 2000, 13, 230-232.

[38] Kovsh, A. R.; Egorov, A. Yu.; Zhukov, A. E.; Ustinov, V. M.; Shernyakov, Yu. M.; Maximov, M. V.; Tsatsulínikov, A. F.; Volovik, B. V.; Lunev, A. V.; Ledentsov, N. N.; Kop'ev, P. S.; Alferov, Zh. I.; Bimberg, D. *Semiconductors*. 1998, 32, 997-1000.

[39] Ustinov, V. M.; Zhukov, A. E.; Egorov, A. Yu.; Kovsh, A. R.; Zaitsev, S. V.; Gordeev, N. Yu.; Kopchatov, V. I.; Ledentsov, N. N.; Tsatsul'nikov, A. F.; Volovik, B. V.; Kop'ev, P. S.; Alferov, Zh. I.; Ruvimov, S. S.; Liliental-Weber, Z.; Bimberg, D. *Electron. Lett.* 1998, 34, 670-672.

[40] Kovsh, A. R.; Zhukov, A. E.; Egorov, A. Yu.; Ustinov, V. M.; Ledentsov, N. N.; Maximov, M. V.; Tsatsul'nikov, A. F.; Kop'ev, P. S. *Semiconductors*. 1999, 33, 184-191.

[41] Zhukov, A. E.; Kovsh, A. R.; Maleev, N. A.; Mikhrin, S. S.; Ustinov, V. M.; Tsatsul'nikov, A. F.; Maximov, M. V.; Volovik, B. V.; Bedarev, D. A.; Shernyakov, Yu. M.; Kop'ev, P. S.; Alferov, Zh. I.; Ledenstov, N. N.; Bimberg, D. *Appl. Phys. Lett.* 1999, 75), 1926-1928.

[42] Huang, X.; Stinz, A.; Hains, C. P.; Liu, G. T.; Cheng, J.; Malloy, K. J. *Electron. Lett.* 2000, 36, 41-42.

[43] Mikhrin, S. S.; Zhukov, A. E.; Kovsh, A. R.; Maleev, N. A.; Vasil'ev, A. P.; Semenova, E. S.; Ustinov, V. M.; Kulagina, M. M.; Nikitina, E. V.; Soshnikov, I. P.; Shernyakov, Yu. M.; Livshits, D. A.; Kryjanovskaya, N. V.; Sizov, D. S.; Maximov, M. V.; Tsatsul'nikov, A. F.; Ledentsov, N. N.; Bimberg, D.; Alferov, Zh. I. *Semiconductors*. 2002, 36, 1315-1321.

[44] Zhukov, A. E.; Kovsh, A. R.; Mikhrin, S. S.; Vasil'ev, A. P.; Semenova, E. S.; Maleev, N. A.; Ustinov, V. M.; Kulagina, M. M.; Nikitina, E. V.; Soshnikov, I. P.; Shernyakov, Yu. M.; Livshits, D. A.; Kryjanovskaya, N. V.; Sizov, D. S.; Maximov, M. V.; Tsatsul'nikov, A. F.; Ledentsov, N. N.; Bimberg, D.; Alferov, Zh. I. *Physica E*. 2003, 17, 589-592.

[45] Kovsh, A. R.; Maleev, N. A.; Zhukov, A. E.; Mikhrin, S. S.; Vasil'ev, A. P.; Shernyakov, Yu. M.; Maximov, M. V.; Livhsits, D. A.; Ustinov, V. M.; Alefrov, Zh. I.; Ledentsov, N. N.; Bimberg, D. *Electron. Lett.* 2002, 38, 1104-1106.

[46] Klopf, F.; Reithmaier, J. P.; Forchel, A. *Appl. Phys. Lett.* 2000, 77, 1419-421.

[47] Zhukov, A. E.; Kovsh, A. R.; Mikhrin, S. S.; Maleev, N. A.; Ustinov, V. M.; Livshits, D. A.; Tarasov, I. S.; Bedarev, D. A.; Maximov, M. V.; Tsatsul'nkov, A. F.; Soshnikov, I. P.; Kop'ev, P. S.; Alferov, Zh. I.; Ledentsov, N. N.; Bimberg, D. *Electron. Lett.* 1999, 35, 1845-1847.

[48] Zhukov, A. E.; Kovsh, A. R.; Livshits, D. A.; Ustinov V. M.; Alferov, Zh. I. *Semicond. Sci. Technol.* 2003, 18, 774-781.

[49] Smowton P. M.; Blood, P. *IEEE J. Quantum Electron.* 1997, 3, 491-498.

[50] Liu, G. T.; Stinz, A.; Li, H.; Newell, T. C.; Gray, A. L.; Varangis, P. M.; Malloy, K. J.; Lester, L. F. *IEEE J. Quantum Electron.* 2000, 36, 1272-1279.

[51] Bimberg, D.; Ledentsov, N. N.; Kirstaedter, N.; Schmidt, O.; Grundmann, M.; Ustinov, V. M.; Egorov, A. Yu.; Zhukov, A. E.; Maximov, M. V.; Kop`ev, P. S.; Alferov, Zh. I.; Ruvimov, S. S.; Gosele, U.; Heydenreich, J. *Jpn. J. Appl. Phys. B.* 1996, 35, 1311-1319.

[52] Zhukov, A. E.; Kovsh, A. R.; Ustinov, V. M.; Shernyakov, Yu. M.; Mikhrin, S. S.; Maleev, N. A.; Kondrat'eva, E. Yu.; Livshits, D. A.; Maximov, M. V.; Volovik, B. V.; Bedarev, D. A.; Musikhin, Yu. G.; Ledentsov, N. N.; Kop'ev, P. S.; Alferov, Zh. I.; Bimberg, D. *IEEE Photon. Technol. Lett.* 1999, 11, 1345-1347.

[53] Ledentsov, N. N.; Kovsh, A. R.; Shchukin, V. A.; Mikhrin, S. S.; Krestnikov, I. L.; Kozhukhov, A. V.; Karachinsky, L. Ya.; Maximov, M. V.; Novikov, I. I.; Shernyakov, Yu. M.; Soshnikov, I. P.; Zhukov, A. E.; Portnoi, E. L.; Ustinov, V. M.; Gerthsen, D.; Bhattacharya, P.; Zakharov, N. F.; Werner, P.; Hopfer, F.; Kuntz M.; Bimberg, D. *Proc. SPIE.* 2004, 6524-01.

[54] Lundina, E. Yu.; Shernyakov, Yu. M.; Maksimov, M. V.; Kayander, I. N.; Tsatsul'nikov, A. F.; Ledentsov, N. N.; Zhukov, A. E.; Maleev, N. A.; Mikhrin, S. S.; Ustinov, V. M.; Alferov, Zh. I.; Bimberg, D. *Tech. Phys.* 2003, 48, 131-132.

[55] Maximov, M. V.; Shernyakov, Yu. M.; Kaiander, I. N.; Bedarev, D. A.; Kondrat'eva, E. Yu.; Kop'ev, P. S.; Kovsh, A. R.; Maleev, N. A.; Mikhrin, S. S.; Tsatsul'nikov, A. F.; Ustinov, V. M.; Volovik, B. V.; Zhukov, A. E.; Alferov, Zh. I.; Ledentsov, N. N.; Bimberg, D. *Electron. Lett.* 1999, 35, 2038-2039.

[56] Mikhrin, S. S.; Zhukov, A. E.; Kovsh, A. R.; Maleev, N. A.; Ustinov, V. M.; Shernyakov, Yu. M.; Kayander, I. N.; Kondrat'eva, E. Yu.; Livshits, D. A.; Tarasov, I. S.; Maksimov, M. V.; Tsatsul'nikov, A. F.; Ledentsov, N. N.; Kop'ev, P. S.; Bimberg, D.; Alferov, Zh. I. *Semiconductors.* 2000, 34, 119-121.

[57] Kovsh, A. R.; Ledentsov, N. N.; Zhukov, A. E.; Livshits, D. A.; Maleev, N. A.; Maximov, M. V.; Ustinov, V. M.; Wang, J.-S.; Chi, J. Y.; Ouyang, D. N.; Bimberg, D.; Lott, J. A. *Proc. SPIE.* 2004, 5349-06.

[58] Temkin, H.; Coblentz, D.; Logan, R. A.; van der Ziel, J. P.; Tanbun-Ek, T.; Yadvish, R. D.; Sergent, A. M. *Appl. Phys. Lett.* 1993, 62, 2402-2404.

[59] Zhukov, A. E.; Ustinov, V. M.; Egorov, A. Yu.; Kovsh, A. R.; Tsatsul`nikov, A. F.; Ledentsov, N. N.; Zaitsev, S. V.; Gordeev, N. Yu.; Kop`ev, P. S.; Alferov, Zh. I. *Jpn. J. Appl. Phys.* 1997, 36, 4216-4218.

[60] Sugawara, M.; Mukai, K.; Nakata, Y. *Appl. Phys. Lett.* 1999, 74, 1561-1563.

[61] Shchekin O. B.; Deppe, D. G. *Appl. Phys. Lett.* 2002, 80, 2758-2760.

[62] Shchekin, O. B.; Ahn, J.; Deppe, D. G. *Electron. Lett.* 2002, 38, 712-713.

[63] Shchekin O. B.; Deppe, D. G., *Appl. Phys. Lett.* 2002, 80, 3277-3279.

[64] Bhattacharya, P.; Mi, Z.; Fathpour, S.; Kovsh, A. R.; Mikhrin, S. S.; Krestnikov, I. L.; Ledentsov, N. N., *Electron. Lett.* to be published.

[65] Hatori, N.; Otsubo, K.; Ishida, M.; Akiyama, T.; Nakata, Y.; Ebe, H.; Okumura, Sh.; Yamamoto, T.; Sugawara, M.; Arakwa, Y. *30th European Conf. Optical Commun.* September 5-9, 2004, Stockholm, Sweden.

[66] Tan, K. T.; Marinelli, C.; Thompson, M. G.; Wonfor, A.; Silver, M.; Sellin, R. L.; Penty, R. V.; White, I. H.; Kuntz, M.; Lammlin, M.; Ledentsov, N. N.; Bimberg, D.; Zhukov, A. E.; Ustinov, V. M.; Kovsh, A. R. *IEEE Photon. Technol. Lett.* 2004, 16, 1415-1417.

[67] Kosogov, A. O.; Werner, P.; Gosele, U.; Ledentsov, N.N.; Bimberg, D.; Ustinov, V. M.; Egorov, A. Y.; Zhukov, A. E.; Kopev, P. S.; Bert, N. A.; Alferov Z. I. *Appl. Phys. Lett.* 1996, 69, 3072-3074.

[68] Nikitina, E. V.; Zhukov, A. E.; Vasil'ev, A. P.; Semenova, E. S.; Gladyshev, A. G.; Kryzhanovskaya, N. V.; Maksimov, M. V.; Shernyakov, Yu. M.; Ustinov, V. M.; Ledentsov, N. N. *Tech. Phys. Lett.* 2004, 30, 644-646.

[69] Ouyang, D.; Ledentsov, N. N.; Bimberg, D.; Kovsh, A. R.; Zhukov, A. E.; Mikhrin S.S.; Ustinov, V. M. *Semicond. Sci. Technol.* 2003, 18, L53-L54.

[70] Maleev, N. A.; Egorov, A. Yu.; Zhukov, A. E.; Kovsh, A. R.; Vasil'ev, A. P.; Ustinov, V. M.; Ledentsov, N. N.; Alferov, Zh. I. *Semiconductors.* 2001, 35, 847-853.

[71] Babic, D. I.; Piprek, J.; Streubel, K.; Mirin, R. P.; Margalit, N. M.; Mars, D. E.; Bowers, J. E.; Hu, E. L. *IEEE J. Quantum. Electron.* 1997, 33, 1369-1383.

[72] Huffaker, D. L.; Deppe, D. G. *IEEE Photon. Technol. Lett.* 1999, 11, 934-936.

[73] MacDougal, M. H.; Dapkus, P. D.; Bond, A. E.; Lin, C.-K.; Geske, J. *IEEE J. Selected Topics Quantum Electron.* 1997, 3, 905-915.

[74] Lott, J. A.; Ledentsov, N. N.; Ustinov, V. M.; Maleev, N. A.; Zhukov, A. E.; Kovsh, A. R.; Maximov, M. V.; Volovik, B. V.; Alferov, Zh. I.; Bimberg, D. *Electron. Lett.* 2000, 36, 1384-1385.

[75] Ustinov, V. M.; Zhukov, A. E.; Maleev, N. A.; Kovsh, A. R.; Mikhrin, S. S.; Cherkashin, N. A.; Shernyakov, Yu. M.; Maximov, M. V.; Tsatsul'nikov, A. F.; Ledentsov, N. N.; Alferov, Zh. I.; Loot, J. A.; Bimberg, D. *Proc. SPIE.* 2002, 4646-05.

[76] Lott, J. A.; Ledentsov, N. N.; Kovsh, A. R.; Ustinov V. M.; Bimberg, D. *Proc IEEE LEOS.* 2003, 499–500.

[77] Maximov, M. V.; Volovik, B. V.; Sotomayor Torres, C. M.; Ramushina, E. M.; Scopina, V. I.; Tanklevskaya, E. M.; Gurevich, S. A.; Ustinov, V. M.; Alferov, Zh. I.; Ledentsov, N. N.; Bimberg, D. *Phys. Stat. Sol (a).* 2001, 188, 955-959.

[78] Ledentsov, N. N.; Maximov, M. V.; Bimberg, D.; Maka, T.; Sotomayor Torres, C. M.; Kochnev, I. V.; Krestnikov, I. L.; Lantratov, V. M.; Cherkashin, N. A.; Musikhin Yu. M.; Alferov, Zh. I. *Semicond. Sci. Technol.* 2000, 15, 604-607.

[79] Ustinov, V. M.; Weber, E. R.; Ruvimov, S.; Liliental-Weber, Z.; Zhukov, A. E.; Egorov, A. Yu.; Kovsh, A. R.; Tsatsul'nikov, A. F.; Kop'ev, P. S. *Appl. Phys. Lett.* 1998, 72, 362-364.

[80] Zhukov, A. E.; Vasil'ev, A. P.; Kovsh, A. R.; Mikhrin, S. S.; Semenova, E. S.; Egorov, A. Yu.; Odnoblyudov, V. A.; Maleev, N. A.; Nikitina, E. V.; Kryzhanovskaya, N. V.; Gladyshev, A. G.; Shernyakov, Yu. M.; Maximov, M. V.; Ledentsov, N. N.; Ustinov, V. M.; Alferov, Zh. I. *Semiconductors*. 2003, **37**, 1461-1464.

[81] Gozu, Sh.-I.; Tsuboki, K.; Hayashi, M.; Hong, Ch.; Yamada, S. *J. Cryst. Growth*. 1999, 201/202, 749-752.

[82] Zhukov, A. E.; Kovsh, A. R.; Mikhrin, S. S.; Semenova, E. S.; Maleev, N. A.; Vasil'ev, A. P.; Nikitina, E. V.; Kryzhanovskaya, N. V.; Gladyshev, A. G.; Shernyakov, Yu. M.; Musikhin, Yu. G.; Maksimov, M. V.; Ledentsov, N. N.; Ustinov, V. M.; Alferov, Zh. I. *Semiconductors*. 2003, 37, 1119-1122.

[83] Ledentsov, N. N.; Kovsh, A. R.; Zhukov, A. E.; Maleev, N. A.; Mikhrin, S. S.; Vasil'ev, A. P.; Semenova, E. S.; Maximov, M. V.; Shernyakov, Yu. M.; Kryzhanovskaya, N.V.; Ustinov, V. M.; Bimberg, D. *Electron. Lett.* 2003, 39, 1126-1128.

[84] Zhukov, A.; Vasiljev, A.; Semenova, E.; Kryzhanovskaya, N.; Gladyshev, A.; Maximov, M.; Ustinov, V.; Ledentsov, N. *Proc. 4th Int. Conf. Advanced Optical Materials Devices (AOMD-4)*, Tartu, Estonia, July 6-9, 2004; Abstract book, p. 10.

[85] Ha, W.; Gambin, V.; Bank, S.; Wistey, M.; Yuen, H.; Kim, S.; Harris, Jr., J. S. *IEEE J. Quantum Electron.* 2002, 38, 1260-1267.

In: Frontiers in Quantum Dots Research
Editor: Marco L. Voronin, pp. 51-67
ISBN: 1-60021-057-0

Chapter 2

SEMICONDUCTOR QUANTUM DOTS BASED ON SPIN-ORBIT INTERACTION

Manuel Valín-Rodríguez*
Departament de Física, Universitat de les Illes Balears,
E-07122 Palma de Mallorca, Spain

Introduction

In the last two decades, the field concerning quantum dots, and, in general, low dimensional semiconductor structures has experienced great advances from both experimental and theoretical points of view. At present, these structures constitute one of the most suitable scenarios to investigate quantum phenomena in condensed matter. In particular, zero-dimensional structures (quantum dots) provide an ideal system to achieve this purpose due to the intrinsic quantum nature of the dynamics characterizing the particles confined; for instance, shell structure of the quantized levels constitutes an illustrative example and the origin of the terminology 'artificial atoms' because of the similarity with natural atoms [1].

Nowadays, a great variety of quantum dots can be fabricated due to the different growing tecniques. Despite of this variety, the origin of the interaction responsible of the confinement has an electrostatic nature; in some cases the mismatch of the band structure of different materials provides the energy step necessary to generate the confinement, in other cases the energy gradient is obtained by directly generating a potential landscape through external electrodes [1].

More recently, another fundamental quantum property of matter, the spin, has attracted a lot of attention in the context of low-dimensional systems. There exist proposals of dots based purely on this property; these proposals rely on a spin-dependent energy gradient to create a confinement that is only effective for a particular spin collectivity. This spin-selectivity is obtained through a spatially inhomogeneous magnetic interaction, such as the Zeeman energy corresponding to an inhomogeneous magnetic field. In the case of GaAs-based dots it has

* E-mail address: vdfsmvr4@clust.uib.es; Tel.: +34 971 17 25 38; Fax: +34 971 17 34 26

been proposed to take advantadge of the high polarizability of the nuclei forming the underlying lattice to create such a magnetic field [2,3].

In this chapter we discuss, within a theoretical model, the feasibility of a novel class of quantum dots not based in any magnetic interaction, where the spin of the particles plays a relevant role determining the confining properties of the dot. The idea is to take advantadge of the intrinsic interactions involving the spin of the carriers present in semiconductor heterostructures instead of using ferromagnetic contacts, magnetic impurities or any source of external magnetic fields.

The main interaction concerning the spin in non-magnetic semiconductors is the spin-orbit coupling; this interaction stems from a relativistic effect caused by the electric fields present in the semiconductor heterostructure where the particles are confined. The presence of electric fields makes the 'relativistic' particles feel an effective magnetic field when transformed to their intrinsic reference frame which induces the spin differentiation. In particular, III-V semiconductor alloys, object of the present work, show two different sources of spin-orbit coupling. Depending on the origin of the electric field causing the effect the two sources are distinguished. On one hand, the lack of inversion symmetry of the bulk material gives rise to an electric field origin of the so-called Dresselhaus spin-orbit term [4]. On the other hand, the electric field caused by the asymmetry present in certain heterostructures due to their particular fabrication settings (structural asymmetry) is the origin of the Bychkov-Rashba term [5]. The latter is characterized by a linear dependence on the momentum of the particles and it has been shown experimentally by several groups that its intensity can be tuned externally by applying to the heterostructure an electric field perpendicular to the plane of the quantum well [6,7,8]. Due to its controllability this spin-orbit mechanism will constitute the basis of the quantum dots proposed in this work.

In the case of the Dresselhaus term, the dependence on the momentum of the particles is cubic and several terms obtained by ciclic permutation of the momentum operators appear, as we will discuss later the intensity of this mechanism depends mainly on the effective width of the two-dimensional electron gas (2DEG) contained in the heterostructure.

As mentioned above, the intensity of the Bychkov-Rashba interaction depends on the vertical electric field created by the structural asymmetry, the degree of asymmetry depends on the specific fabrication settings of each particular heterostructure; as a consequence, the intensity of the Bychkov-Rashba term is characterized by a noticeable sample-dependence. Note that, this dependence on the fabrication process supposes that, in principle, the intensity of the spin-orbit could be engineered. Furthermore, as mentioned above, the Bychkov-Rashba coupling can be controlled *in situ* by means of external electrodes. These properties of the Bychkov-Rashba interaction provide the basical requirements in order to create the proposed quantum dots, since they rely on the possibility of spatial control of the spin-orbit interaction, i.e., a spatial-modulation of the spin-orbit intensity is necessary to achieve the desired confining effect. Actually, the microscopic inhomogeneity of the heterostructure's vertical electric field makes the Bychkov-Rashba intensity intrinsically space-dependent for certain types of heterostructures [9,10].

The most straightforward solution to the problem of creating a space-modulated spin-orbit intensity is the usage of 'local' electrodes that would externally modify the vertical electric field in the region of the heterostructure covered by the electrodes, thus controlling the magnitude of the spin-orbit modulation and the spatial region where this holds [11,12]. The main problem of this approach is that the usage of local electrodes would not only modify the 'effective' vertical electric field present in the heterostructure, as it could also generate in-plane gradients of electric potential leading to hybrid quantum dots where both mechanisms are present: the usual electrostatic in-plane fields and the spin-orbit intensity gradient.

A deeper insight in the factors determining the magnitude of the Bychkov-Rashba spin-orbit intensity is required in order to obtain a more refined method to achieve the desired spatial modulation. In this respect Andrada et al. have shown the relevance of the barrier layers in the effects caused by the spin-orbit interaction for several types of semiconductor heterostructures using a multiband '$k \cdot p$' model [13]. More recently, Grundler, using an InAs-based quantum well with InGaAs barrier layers, obtained experimentally large values of spin-orbit intensity that cannot be explained with multiband $k \cdot p$ predictions. Furthermore, this author showed that this sample was characterized by a large range of tunability of the spin-orbit intensity (of the order of the zero-bias value) by applying externally a vertical electric field [8]. An intriguing fact is that the applied electric fields were of the order of 10^3 V/cm, while the built-in electric field present in the sample was of the order of 10^5 V/cm. In view of this, a relevant question arises: why a relatively weak external electric field can lead to sizeable variations of the zero-bias spin-orbit intensity?. The answer of this question lies in the role played by the barrier layers; Grundler argues that the main effect of the applied field consists in a displacement of the profile of the wavefunction perpendicular to the well's plane. The wavefunction penetrates into the barrier layer while the modification of the total vertical field is very small. This implies that the penetration in the barrier layers constitutes a very relevant parameter when determining the Bychkov-Rashba spin-orbit intensity.

The strong dependence of the spin-orbit intensity on the wavefunction's penetration into the barrier layers constitutes a very advantadgeous property in order to create a system characterized by a space-engineered spin-orbit interaction, since the usage of properly inhomogenized barrier layers would lead to the fabrication of delimited regions within the heterostructure, characterized by different spin-orbit intensities depending on the 'local' properties of the barriers. One way to produce the inhomogeneity in the barrier layers is given by the usage of different materials in their spatial composition: in this way, a zone with high-barrier material will experience lower penetration of the wavefunction than in the remaining barrier layer. Another possibility to control the wavefunction's penetration comes from the usage of barriers with inhomogeneous thickness; depending on the properties of the substrate material, the penetration of the wavefunction would be enhanced or lowered in the regions of the heterostructure characterized by different thickness.

One disadvantadge of this approach with respect to the electrodes-based one is that the tunability property is lost; once the inhomogeneity of the barrier layers is fixed by the fabrication process the inhomogeneity in the magnitude of the spin-orbit intensity keeps also fixed.

Modelization

Once the basis of the confining mechanism necessary to create the proposed quantum dots has been established, we present the theoretical model used to quantify their feasibility.

In particular, we are interested in electronic dots, i.e., the particles spatially confined are conduction band electrons. To characterize these particles we use the effective mass approximation for the conduction band of the semiconductor. This approximation works rather well in the case of certain wide-gap semiconductors, such as GaAs, but, in the case of narrow gap semiconductors, for instance InAs, the proximity of the different energy bands supposes that the interplay between them becomes important and the effective mass approximation loses applicability. In this sense, a more suitable model to achieve a quantitative description of the quantum dots is given by multiband '$k \cdot p$' models taking into account, basically, the coupling between the conduction and the valence bands [14]. As we will not restrict our treatment to a particular semiconductor material the approximations of our model might be quite drastic in some cases and, therefore, the scope of this chapter is limited to a qualitative discussion of the feasibility rather than to a detailed quantitative treatment.

The proposed dots are based in a III-V heterostructure containing a 2DEG; for the typical sizes of the dots that we will consider (exceeding a characteristic radius of 100 nm) the system can still be considered as a two-dimensional one. Any heterostructure constitutes a three-dimensional system, with the peculiarity that the orbital degree of freedom perpendicular to the heterostructure's plane is strongly quantized, as the typical scale of movement in this direction is of the order of a few nanometers. A quantum dot characterized by typical sizes of a fraction of a micrometer defined within a heterostructure can still be considered as two-dimensional, since a change of state in the perpendicular direction involves a great amount of energy in comparison with the energy concerning the in-plane state transitions. Therefore, the perpendicular motion can be considered as effectively 'frozen'and decoupled from the in-plane motion.

In this two-dimensional representation, the Hamiltonian characterizing the motion within the dot is only composed of the kinetic energy and the spin-orbit interaction:

$$H_0 = \frac{1}{2m^*}\left(p_x^2 + p_y^2\right)$$

$$H_D = \frac{\lambda_D}{\hbar}\left(p_x\sigma_x - p_y\sigma_y\right)$$

$$H_R = \frac{\lambda_R}{\hbar}\left(p_y\sigma_x - p_x\sigma_y\right)$$

$$H = H_0 + H_D + H_R$$

where p_x and p_y represent the components of the in-plane momentum of the electron, m^* represents the effective mass corresponding to the bottom of the conduction band, the σ's denote the Pauli matrices corresponding to the in-plane spin orientations and λ_D, λ_R correspond to the Dresselhaus and Bychkov-Rashba spin-orbit intensities, respectively. These two quantities are usually expressed in terms of intrinsic parameters of the heterostructure; for instance, the main contribution of the Dresselhaus interaction is given by a term containing a cuadratic dependence on the momentum perpendicular to the heterostructure's plane 'k_z', since the modelization is two-dimensional the third dimension is parametrized through the effective width of the 2DEG 'z_0', usually, this intensity is expressed as $\lambda_D = \gamma(\pi / z_0)^2$ where γ corresponds to a material especific constant which strongly depends on the semiconductor alloy considered [15]. In the case of the Bychkov-Rashba interaction the parameter used is the 'effective' vertical electric field, within this characterization λ_R is linear in this 'effective' field (E_0) and also proportional to a material especific constant 'α' ($\lambda_R = \alpha e E_0$) having also a rather strong dependence on the alloy considered [15].

As stated above, the three-dimensional structure is needed in order to include the spatial modulation of the spin-orbit interaction, since the three-dimensional structure (role of the barrier layers) is the real responsible of the inhomogeneity leading to the desired spatial modulation. As we are restricted to a two-dimensional model we directly include the inhomogeneity in the three-dimensional structure through a spatial dependence in the spin-orbit parameter λ_R.

The scope of this work is limited to circular quantum dots corresponding to a spatial dependence of the spin-orbit parameter characterized by circular symmetry,

$$\lambda_R(x, y) = \lambda_R(r)$$

where 'r' and 'θ' are the usual in-plane polar coordinates. This symmetry will allow us to make some analytical considerations in the case where only one source of spin-orbit interaction (Bychkov-Rashba term) is present. The particular radial dependence chosen to model the quantum dots is given by a step-like function,

$$\lambda_R(r) = \lambda_e + (\lambda_i - \lambda_e)\frac{1}{1 + e^{(r-R_0)/\sigma}}$$

where λ_i represents the spin-orbit intensity within the region of the dot and λ_e represents the spin-orbit intensity in the bulk surrounding region. The nominal radius of the quantum dot is given by R_0 and σ is a small diffusivity introduced to avoid discontinuities in the numerical solutions.

Despite of the radial symmetry of the Bychkov-Rashba intensity the whole spin-orbit term is not rotationally invariant. However, a symmetry combining orbital and spin degrees of freedom still holds. Consider the case where only the Bychkov-Rashba interaction is present, under this condition the full Hamiltonian is invariant under a combined rotation in real and spin spaces by the same angle and sense only if the intensity is a purely radial function, i.e., the Hamiltonian conmutes with the generalized angular momentum $J_z = L_z + S_z$. Note that, at variance with conventional circular quantum dots, the Hamiltonian no longer conmutes with L_z and S_z separately.

The opposite case, when the Dresselhaus interaction is the only spin-orbit source present, the Hamiltonian is still invariant under a combined rotation in real and spin spaces, however, in this case, rotations in real and spin spaces have the same magnitude but different sense, i.e., the Hamiltonian conmutes with the operator $O_Z = L_Z - S_Z$. The validity of this statement is conditioned again to a pure radial dependence of the spin-orbit intensity.

In this chapter, we will only deal with the Bychkov-Rashba interaction with radial-dependent intensity. In this case, as a consequence of the J_Z-symmetry of the single-particle Hamiltonian, the eigenfunctions can be expressed, in a general form, as a linear superposition of those corresponding to the J_Z operator having the same eigenvalue 'j', i.e.,

$$e^{il\theta}\begin{pmatrix}1\\0\end{pmatrix},\; e^{i(l+1)\theta}\begin{pmatrix}0\\1\end{pmatrix} \Rightarrow j = l + 1/2$$

where 'l' corresponds to an angular-momentum-like index, 'θ ' is the azimutal polar coordinate and the spinorial parts are represented in the usual σ_Z-basis. The coeficients of this superposition depend on the radial coordinate and will yield the radial properties of the Hamiltonian's solutions.

$$\Psi_{nj}(r,\theta) = \begin{pmatrix}\phi_{nl\uparrow}(r)e^{il\theta}\\ \phi_{nl\downarrow}(r)e^{i(l+1)\theta}\end{pmatrix} = \begin{pmatrix}\phi_{nj\uparrow}(r)e^{i(j-1/2)\theta}\\ \phi_{nj\downarrow}(r)e^{i(j+1/2)\theta}\end{pmatrix}$$

where (n,j) and (n,l) constitute equivalent sets of quantum numbers characterizing the states; $n = 0,1,2,\ldots$, $j = \pm 1/2, \pm 3/2, \ldots$, $l = 0, \pm 1, \pm 2, \ldots$ Taking advantadge of this symmetry characterizing the Hamiltonian, the azimutal dependence of the eigenfunctions can be factorized and the two-dimensional problem is reduced to a pair of one-dimensional coupled radial equations for the two spinor components,

$$-\frac{\hbar^2}{2m^*}\left[\frac{\partial\phi_\uparrow}{\partial r^2} + \frac{1}{r}\frac{\partial\phi_\uparrow}{\partial r} - \frac{(j-1/2)^2}{r^2}\phi_\uparrow\right] - \lambda_R(r)\left[\frac{\partial\phi_\downarrow}{\partial r} + \frac{j+1/2}{r}\phi_\downarrow\right] = \varepsilon_{nj}\phi_\uparrow$$

$$-\frac{\hbar^2}{2m^*}\left[\frac{\partial\phi_\downarrow}{\partial r^2} + \frac{1}{r}\frac{\partial\phi_\downarrow}{\partial r} - \frac{(j+1/2)^2}{r^2}\phi_\downarrow\right] + \lambda_R(r)\left[\frac{\partial\phi_\uparrow}{\partial r} - \frac{j-1/2}{r}\phi_\uparrow\right] = \varepsilon_{nj}\phi_\downarrow$$

where $\phi_\uparrow$ and $\phi_\downarrow$ denote $\phi_{nj\uparrow}(r)$ and $\phi_{nj\downarrow}(r)$, respectively.

In general, the above model has to be solved numerically; in this sense we have used two alternative numerical procedures to analyze the model. The first method consists in the resolution of the above coupled radial equations in an uniform one-dimensional grid representing the radial coordinate, whereas the second method consists in solving the initial two-dimensional problem without imposing any symmetry restriction in real neither spin space; the resolution is based in the Cartesian representation of the Hamiltonian which is implemented in an uniform two-dimensional grid. The latter procedure is the only one suitable in the case that no symmetry holds in the Hamiltonian; for instance, when both sources of spin-orbit are present or when the spatial dependence of the spin-orbit intensity is not radial. To prove the reliability of the results obtained from the simulations we have checked that both procedures yield the same results with high accuracy.

Before presenting the numerical results, let us consider some properties characterizing the above Hamiltonian. First of all, it has to be noted that any system including the spin-orbit interaction is still invariant under time reversal. Consequently, a two-fold degeneracy, known as Kramers degeneracy, must hold. From the radial equations it can easily be seen that the substitutions: $\phi_\uparrow \rightarrow \phi_\downarrow$, $\phi_\downarrow \rightarrow -\phi_\uparrow$ and $j \rightarrow -j$ lead to a new state (Kramers conjugate) degenerated in energy with the original state ($\varepsilon_{n,j} = \varepsilon_{n,-j}$) and having opposite generalized angular momentum. Angular momentum and spin operators, as stated above, no longer conmute separately with the Hamiltonian. In conventional circular dots (characterized by a radial confining electric potential) these two operators conmute with the Hamiltonian, so two symmetries hold: angular momentum inversion symmetry and spin inversion symmetry. Consequently, states corresponding to conventional dots are characterized by a four-fold (two-fold) degeneracy depending of their angular momentum ('l') value: non-zero (zero). In the case of the dots object of the present work the two-fold degeneracy holds for all the eigenstates, reflecting a genuine spin-orbit effect.

A careful look on the symmetry substitutions show that the conjugate state has an opposite spin character than that of the original state, i.e., if the original state has a predominant 'up' spin orientation the conjugate one has a predominant 'down' character and viceversa. Something similar happens with the sense of rotation corresponding to each state. Despite of 'l' no longer holds as a good angular-momentum number, it still provides the sign and a good approximated value of $\langle L_z \rangle$; for each eigenspinor its conjugate shows an inverted sense of rotation due to the change of sign in the angular-momentum-like index (the substitution $j \rightarrow -j$ is equivalent to $l \rightarrow -(l+1)$). This implies that each pair of Kramers conjugated states shows opposite spin orientation and opposite sense of rotation; this property of the conjugated states can be exploited in order to create permanent charge and/or spin currents within the dot. In particular, for a general eigenstate of J_z the permanent density and spin currents read,

$$\vec{j}(r) = \frac{\hbar}{m^*}\frac{1}{r}\left[j\rho(r) - \frac{m_z(r)}{2} \right]\hat{\theta}$$

$$\vec{j}_{S_z}(r) = \frac{\hbar}{m^*}\frac{1}{r}\left[jm_z(r) - \frac{\rho(r)}{2} \right]\hat{\theta}$$

where $\rho(\mathrm{r})$ and $m_z(\mathrm{r})$ are defined as the usual particle and spin densities, respectively: $\rho(r) = |\phi_{nj\uparrow}(r)|^2 + |\phi_{nj\downarrow}(r)|^2$, $m_z(r) = |\phi_{nj\uparrow}(r)|^2 - |\phi_{nj\downarrow}(r)|^2$ and the single-particle currents are obtained through the formulas:

$$\vec{j}(\vec{r}) = \frac{\hbar}{2im^*}\left[\psi^*(\vec{r})\nabla\psi(\vec{r}) - \psi(\vec{r})\nabla\psi^*(\vec{r}) \right]$$

$$\vec{j}_{S_z}(\vec{r}) = \frac{\hbar}{2im^*}\left[\psi^*(\vec{r})\nabla\sigma_z\psi(\vec{r}) - \psi(\vec{r})\nabla\sigma_z\psi^*(\vec{r})\right]$$

It is worth to mention that, since the radial functions of the eigenstates are real, they do not contribute to the currents within the dot and the whole current flows in the azimutal direction.

Note that, for a given energy of an electron, its corresponding state is not well-defined by a single J_Z-eigenstate. In general, an energy eigenstate is characterized by a linear combination of the Kramers conjugated states,

$$\Psi_n(r,\theta) = \alpha\begin{pmatrix}\phi_{nj\uparrow}(r)e^{i(j-1/2)\theta}\\ \phi_{nj\downarrow}(r)e^{i(j+1/2)\theta}\end{pmatrix} + \beta\begin{pmatrix}\phi_{nj\downarrow}(r)e^{-i(j+1/2)\theta}\\ -\phi_{nj\uparrow}(r)e^{-i(j-1/2)\theta}\end{pmatrix}$$

For this general state of the particle the radial contribution to the currents vanishes again. There also appear terms corresponding to currents periodic in θ whose contribution is also zero since, as we are interested in the permanent currents generated in the dot, when averaging their contribution in the whole range of azimutal directions they give no net current. As a matter of fact, the permanent spin current does not change when the mixing of states is introduced, i.e., it does not depend on the relative weight of the Kramers doublet. On contrary, the particle current depends on the mixing.

$$\vec{j}(r) = \frac{\hbar}{m^*}\frac{|\alpha|^2 - |\beta|^2}{r}\left[j\rho(r) - \frac{m_z(r)}{2}\right]\hat{\theta} \quad .$$

In the particular case where the relative weight of both Kramers conjugated states is equal, there is no net charge current within the dot while a non-zero spin current flows.

Since this property corresponds to a single-particle Hamiltonian let's consider the most paradigmatic system where this Hamiltonian holds: a one-electron dot. In this case, if the dot is not excited by any mechanism the electron rests in its ground state given by the Kramers doublet $\{(n=1, j=-1/2), (n=1, j=1/2)\}$ yielding a permanent spin current and a possible charge current, as stated above. Note that, in the case of conventional (electrostatic) quantum dots there is no analogous situation, since the lowest energy single-particle states are characterized by zero angular momentum which do not generate any current.

Another interesting property coming from the symmetry properties of the Hamiltonian concern the spin structure of the eigenstates. The pure radial dependence of the vertical spin component has been exposed above when considering the currents within the dot. There also exist an in-plane (horizontal) spin structure characterizing the J_Z-eigenstates that displays a characteristic texture.

$$m_z(r) = \langle\sigma_z\rangle(r) = |\phi_{nj\uparrow}(r)|^2 - |\phi_{nj\downarrow}(r)|^2$$

$$\langle \vec{\sigma}_{//} \rangle (r,\theta) = 2\phi_{nj\downarrow}(r)\phi_{nj\uparrow}(r)\left[\cos(\theta)\hat{x} + \sin(\theta)\hat{y}\right]$$

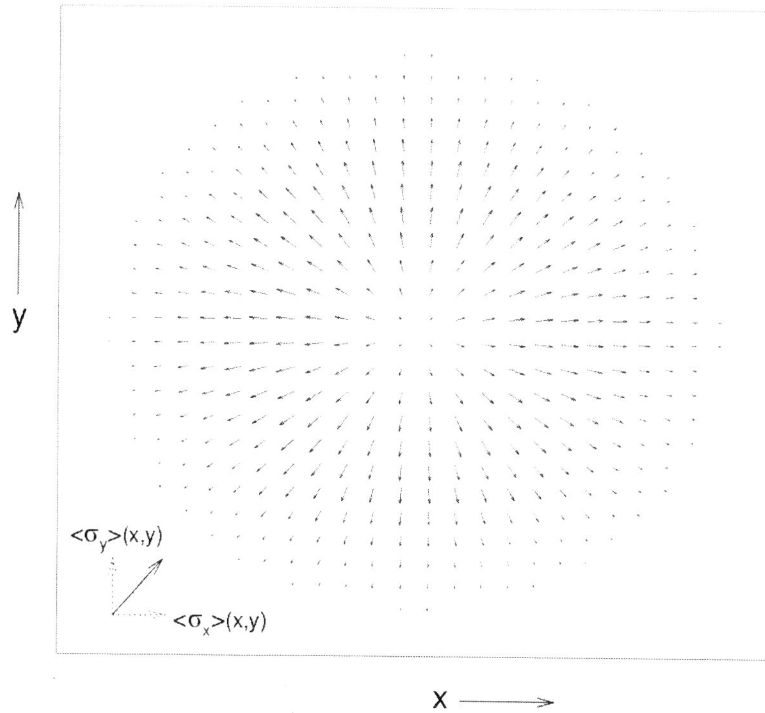

Figure 1. Schematic representation of the spatial distribution corresponding to the in-plane spin texture of an J_z-eigenstate. Each arrow is constructed through to the 'local' expectation values of the in-plane Pauli matrices (see inset).

Note that the in-plane spin distribution always points in a radial direction. This particularity is a direct consequence of the combined rotational invariance in real and spin spaces: consider two points in the dot having the same radius and different azimutal angle 'θ'; due to the symmetry of the Hamiltonian with J_z, the local in-plane spin orientation in the second point must be that of the first point rotated the angle needed to go from the first point to the second in the real space, i.e., in-plane spin rotation must follow the in-plane rotation in real space. This property of the spin-orbit coupling is of particular interest for determining the role of the geometrical phase (Berry's phase) in mesoscopic structures [16,17].

A mathematical feature characterizing the Hamiltonian equations and concerning the spin structure is that, if we consider the coupled equations in terms of the angular-momentum-like label 'l', always one of the two equations looks like a Schrödinger equation for a non-zero angular momentum state. This means that, at least, one of the radial spinorial components will be zero at the origin of coordinates, thus, preventing and undefined in-plane spin texture at the origin. Figure 1 depicts schematically the in-plane spin distribution of the J_z-eigenstates, showing its characteristic radial orientation and a vanishing value at the origin of coordinates.

Numerical Results

The first relevant issue concerning the numerical simulations is the determination of the conditions required to the proposed dots to be confining. In this sense, an important condition that we have obtained from the numerical simulations is that the spin-orbit interaction within the nominal region of the dot must be bigger than that of the surrounding material, i.e., $\lambda_i > \lambda_e$. The reason for this behaviour can be understood from bulk properties: let's consider the bulk energy bands of a two-dimensional system with Bychkov-Rashba spin-orbit interaction:

$$\varepsilon_{k\pm} = \frac{\hbar^2}{2m^*}\left|\vec{k}\right|^2 \pm \lambda\left|\vec{k}\right| = \frac{\hbar^2}{2m^*}\left(\left|\vec{k}\right| \pm \frac{m^*}{\hbar^2}\lambda\right)^2 - \frac{m^*}{2\hbar^2}\lambda^2$$

This energy bands are parabolic in the electronic momentum and they are centered in a $\vec{k}$ -value dependent on the spin-orbit intensity and the spin eigenstate considered. Note also that the bands are shifted respect to the zero-energy by a value proportional to the square of the intensity, this term originates an energy mismatch between systems having different spin-orbit intensities that is the responsible of the confining mechanism. Consider a heterostructure characterized by a high spin-orbit intensity, its energy origin would be quite deeper than that of a system having a lower intensity, consequently, spatial variations of the spin-orbit intensity induce an energy gradient that might lead to a bound system in the region of higher intensity. Note, however, that the confining character of this energy gradient is characterized by a special property; it depends on the spin orientation and the orbital state of the particle. This implies that the confinement has not an usual electrostatic character since it is not based in the charge of the particle.

To illustrate this fact, consider again the bulk bands of two-dimensional systems having different spin-orbit intensities. In order to simplify the argumentation, consider a particular orientation of the electronic momentum; for example $\vec{k} = (k_x, 0)$. In this case, the band structure of both systems are depicted schematically in fig. 2 in terms of the component of the momentum k_x. Note that, only the system characterized by a bigger spin-orbit intensity can achieve lower energies than that having the smaller intensity. However, this statement only holds under certain conditions; for instance, consider an electron whose spin is oriented parallel to any spin eigendirection, namely 'up' ('down'), the energy mismatch 'Δ' can only be confining for the range of negative (positive) momentum 'k_x' as can be seen from fig. 2. This means that this mechanism confines the different spin states depending on the characteristics of their associated orbital motion. In the case of our interest, quantum dots with circular symmetry, this spin-selectivity through the orbital motion can be interpreted as that a certain spin orientation can be confined only if the associated orbital rotation occurs in a certain sense, and viceversa, particles having the opposite spin orientation can only be confined if they rotate in the opposite sense respect to the first ones. In accordance with the derived property that spin currents can be generated within the dot without creating a net charge current.

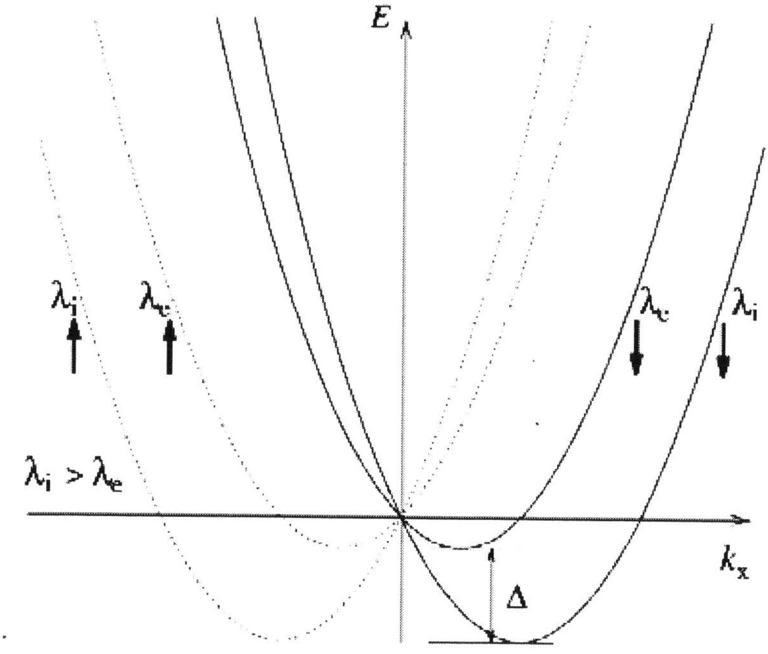

Figure 2. Schematic representation of the energy bands corresponding to a pair of bulk systems with different spin-orbit coupling intensities as a function of one component of the moment.

It is important to mention that these confining properties are originated by the particular form of the spin-orbit Hamiltonian representing the Bychkov-Rashba term. If other form of the spin-orbit Hamiltonian is used; for instance, one including an extra energy offset depending on the spin-orbit intensity, the confining character of the mechanism would be afected. Nevertheless, there exist arguments proving that the Bychkov-Rashba term used has a physical character.

There exist orbital states that do not suffer spin-orbit effects, since the spin-orbit coupling is a relativistic effect, motionless electrons should not suffer this interaction, i.e., states characterized by $\vec{k} = \vec{0}$ should have the same energy independently of the spin-orbit intensity as they do not 'see' this interaction. This can only be accomplished if any term including the spin-orbit intensity in the energy bands is accompanied by a direct proportionality in the momentum. This statement implies that the bands should be given by the formula expressed above

It could be thought that the states which do not suffer the spin-orbit interaction are those whose velocity is zero, where the velocity operator is defined through the equation:

$$\vec{v} = \frac{d\vec{x}}{dt} = -\frac{i}{\hbar}[\vec{p}, H]$$

In this case, zero-velocity states correspond to that located in the minima of the energy bands and the momentum '$\vec{k}$' characterizing them is different from $\vec{k} = \vec{0}$. However, these states do not represent states where spin-orbit effects disappear; namely, if this statement were true, since no other interaction involving the spin is present in the Hamiltonian, the energy of these states should be independent of their associated spin orientation, i.e., the energy of these states should be invariant under a spin rotation. Nevertheless, if we apply a spin rotation to any of these zero-velocity states we observe that their energy vary with the spin orientation; furthermore, the energy variation depends on the spin-orbit intensity. Consequently, it is concluded that zero-velocity states are affected by the spin-orbit interaction. On contrary, states characterized by $\vec{k} = \vec{0}$ satisfy the required property, i.e., their energy is independent of the spin orientation since both spin eigenstates are degenerated for this particular value of the momentum.

Another question of relevance refers to the magnitude of the step in the spin-orbit intensity needed to obtain confined states. Before discussing this issue let us introduce the atomic effective units, appropiate to treat our effective mass Hamiltonian. In this unit system the usual atomic units are modified by the introduction of the parameters characterizing the conduction band and the electrical properties of the semiconductor material, the effective mass (m^*) and the relative dielectric constant (ε_r). The effective units of distance read $a_0^* = \varepsilon_r a_0 / m^*$ where 'a_0' is Bohr radius ($a_0 = 0.529 \cdot 10^{-10}$ m), the unit of energy $H^* = m^* H / \varepsilon_r^2$ where 'H' is the Hartree energy unit (H = 27.2 eV), whereas spin-orbit intensity units, obtained as a product between energy and distance units, read $[\lambda_R] = H^* a_0^* = Ha_0 / \varepsilon_r = 1.43 \cdot 10^{-7} / \varepsilon_r$ eVcm.

Note that, while the value of energy and distance units depend strongly on the material considered through the effective mass, the units of spin-orbit intensity are very homogeneous for the different materials since they only depend on the parameter ε_r, that takes values in the range ε_r = 12-15 for the usual semiconductor alloys. Therefore, a good approximation for this units is given by $[\lambda_R] = 1 \cdot 10^{-8}$ eVcm, independently of the material considered. In this way, the spinorial Schrödinger equation can be written in a dimensionless form leading to results that are to scale with the value of the units.

For a typical size of dot radius of the order of 10 atomic effective units (typically in the range 100-500 nm) simulations show that an intensity range $\lambda_i \approx 1.5 - 2.0 \cdot 10^{-9}$ eVcm is required to obtain single-particle confined states when the external intensity is zero ($\lambda_e = 0$). It is important to specify this external value in order to determine the minimum step needed to achieve bound states, since these two quantities are strongly correlated. For instance, when the external intensity takes a value $\lambda_e \approx 2 \cdot 10^{-9}$ eVcm the required spin-orbit step is reduced and takes a value in the range $\Delta\lambda = \lambda_i - \lambda_e \approx 0.4 - 0.7 \cdot 10^{-9}$ eVcm.

A justification for this reduction of the intensity step when the external value is enhanced can also be obtained from bulk considerations. As stated before, the energy mismatch leading to confinement stems from the different band energy origin of the different regions having different spin-orbit intensities. In a bulk system, this energy origin is proportional to the square of the intensity, therefore, two bulk samples having different intensities '$\lambda_e + \Delta\lambda$' and '$\lambda_e$' are characterized by a characteristic energy mismatch given by:

$$\Delta E \propto (\lambda_e + \Delta\lambda)^2 - \lambda_e^2 = \Delta\lambda(2\lambda_e + \Delta\lambda)$$

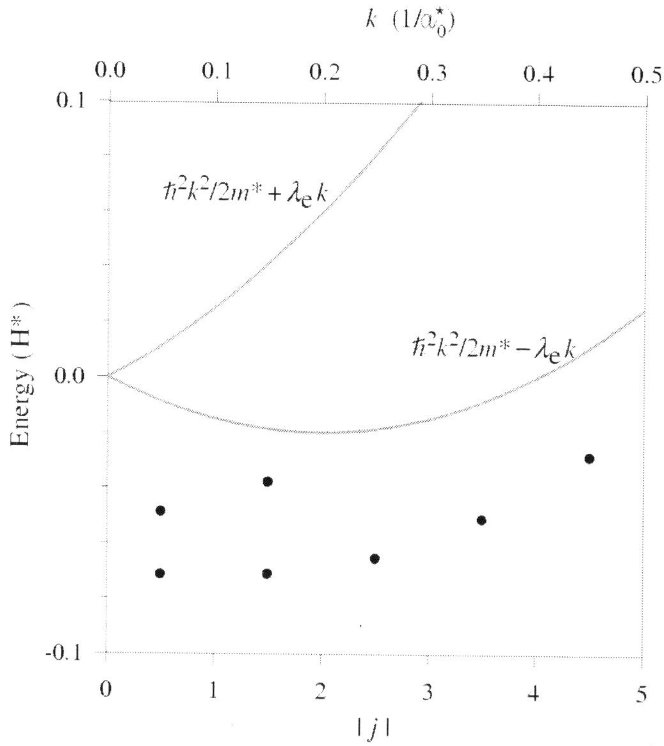

Figure 3. Structure of the bound states as a function of the generalized momentum j for a quantum dot having $\lambda_i = 0.4\ H^*a_0^* \approx 4\cdot10^{-9}$ eVcm, $\lambda_e = 0.2\ H^*a_0^* \approx 2\cdot10^{-9}$ eVcm and a nominal dot radius of 10 atomic effective units (lower scale). There are also plotted the bulk bands (solid lines) corresponding to the surrounding material characterized by λ_e (upper scale).

Once the energy step required to obtain confined states is known, the required spin-orbit step gets reduced as the external intensity (λ_e) is increased, and viceversa, the spin-orbit step required becomes bigger as the external intensity is reduced to achieve the same energy gradient as can be seen from the above formula.

Quantum dots satisfying the above conditions ($\lambda_i > \lambda_e$ and a suitable step $\Delta\lambda$) show a level structure analogous to that of conventional quantum dots. For instance, a dot characterized by spin-orbit parameters in the range of the biggest Rashba intensities experimentally measured [8], $\lambda_i \approx 2\cdot10^{-9}$ and $\lambda_e \approx 2\cdot10^{-9}$, and having a nominal radius of 10 atomic effective units shows a total of 14 single-particle bound states distributed in two subbands, as can be seen in fig. 3. The first one containing 10 states characterized by the quantum numbers n = 1, j = -9/2,-7/2,...,7/2,9/2 and the second one having the remaining 4 states; n = 2, j = -3/2,-1/2,1/2,3/2. There are also depicted the continuum energy bands corresponding to the infinite region outside the dot, which lie at higher energy than the discrete bound states. Figure 4 shows the spatial representation of the lowest energy state of this dot, corresponding to the labels n = 1, j = 1/2, and its conjugated state n = 1, j = -1/2. The upper panels show the radial profile of the spinorial components for both states, the only difference between the two cases are the exchange of the spin index and the sign change in one of the components, in agreement with the symmetry properties derived above. Middle panel shows the spatial distribution of the density which has radial symmetry and is common for both conjugated states. Finally, the lower panel displays the radial distribution of the S_z spin density for the two states, showing that conjugated states posses opposite spin characters.

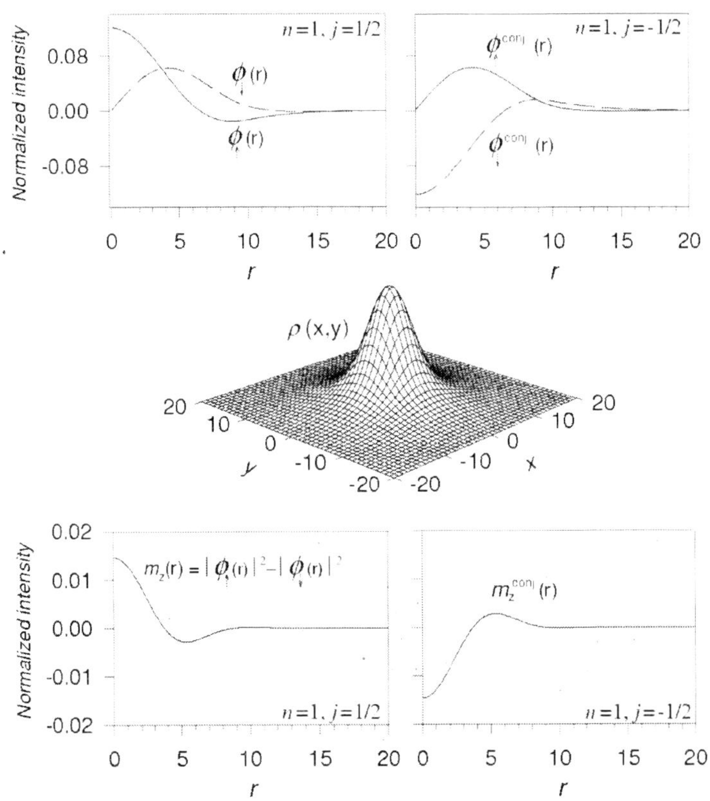

Figure 4. Upper panels: radial spinorial components corresponding to the lowest energy pair of conjugated states of a dot characterized by the same parameters of fig 3. Middle panel: two-dimensional representation of the probability density corresponding to the above Kramers conjugated states. Lower panel: radial profile of the sz spin density distribution corresponding to the same conjugated states. All coordinates are expressed in terms of atomic effective units of distance.

In fig. 5 there are plotted the radial profiles corresponding to the confined states having positive '*j*'of the above dot (negative *j*-states can be obtained through the symmeytry rules). As usually, states of high generalized angular momentum show a more external density distribuiton and the spinorial components corresponding to the highest subband ($n = 2$) show an extra node more than those of the first subband ($n = 1$). Note that, in all cases, the 'up' spinorial component predominates, yielding a spin- density having a net 'up' character; in agreement with the qualitative argument derived from bulk considerations that depending on the characteristics of the orbital motion the mechanism is confining for a certain spin collectivity.

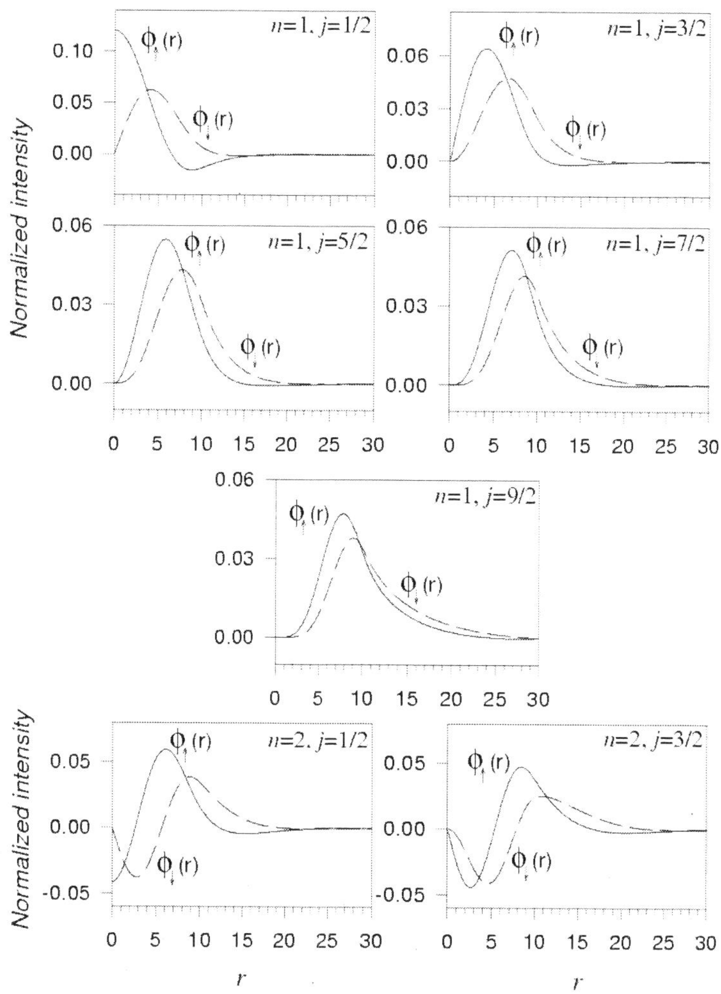

Figure 5. Radial profiles of the spinorial components corresponding to the whole set of bound states represented in fig. 3. In the two lowest panels there are plotted the profiles corresponding to the bound states of the second subband (n = 2), whereas the upper panels represent those corresponding to the first subband (n = 1). All radial coordinates are expressed in terms of atomic effective units.

Up to now, we have only presented results concerning the single-particle properties of the dots. There is also the possibility of the spin-orbit-based confinement to be robust against electron-electron interaction and, thus, capable of supporting multielectron bound states. In order to treat the interacting problem we use the numerical procedure described before based on the implementation of the Hamiltonian in a two-dimensional grid without any symmetry restriction; the electron-electron interaction is included by means of density functional theory. It is used a generalization of the local-spin-density approximation (LSDA) for non-collinear spin densities (see ref. [18] for a previous application of this formalism). Within this scheme, we find that, for reasonable parameters, even two-electron states can be confined to these dots. Upper panel of figure 6 shows the ground state density for an interacting two-electron dot characterized by a nominal radius of 15 units and the same spin-orbit parameters of Fig. 3. Despite of the broken rotational symmetry the LSDA solution shows a symmetric ground state density. Note also that density concentrates in the interface between the regions of different spin-orbit intensity showing a ring-like shape. As an additional check of this result, we have also solved the Hartree-Fock equations for the same parameter set, finding a total energy similar to that of the LSDA solution (discrepancy of about 10 %) and a density confined to the dot. Lower panel of Fig. 6 shows the obtained Hartree-Fock ground state density; within this scheme electrons tend to localize and, consequently, density appears packed in two lobes. Similarly to the LSDA solution, the density appears concentrated again in the interface between the two spin-orbit regions. We have also tried to check the robustness of the mechanism for a three-electron system, but for reasonable parameter values we have not found any bound solution.

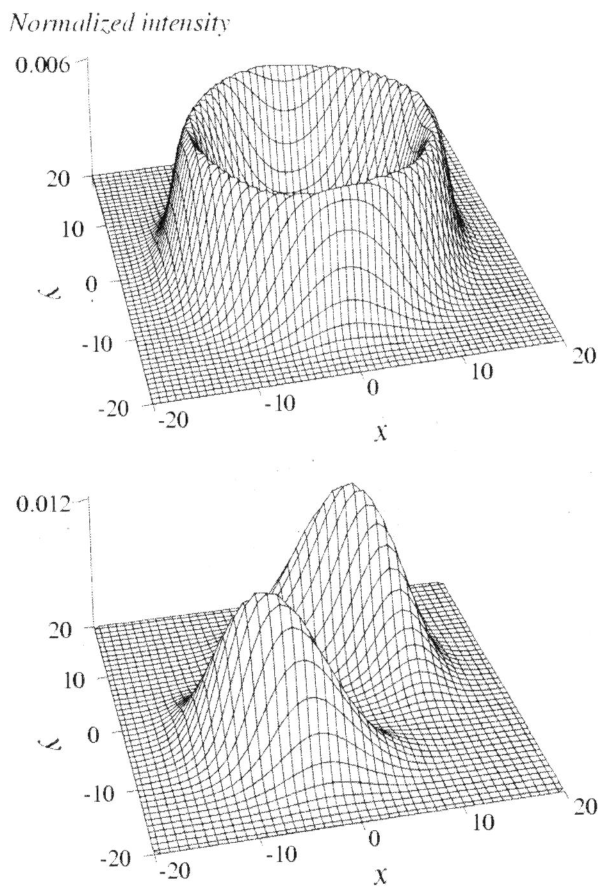

Figure 6. Up: representation of the two-dimensional LSDA density for an interacting two-electron quantum dot having the spin-orbit parameters $\lambda_i = 0.4\ H^*a_0^* \approx 4 \cdot 10^{-9}$ eVcm, $\lambda_e = 0.2\ H^*a_0^* \approx 2 \cdot 10^{-9}$ eVcm and a nominal dot radius of 15 atomic effective units. Down: representation of the Hartree-Fock ground state density for a quantum dot with the same parameter set. Atomic effective units are used to represent the cartesian coordinates.

The introduction of a uniform Dresselhaus spin-orbit term in the model do not substantially modify the confining character of the mechanism. Nevetheless, the presence of both spin-orbit terms destroys the combined symmetry in spin and real spaces, so no analytical properties neither dimensional reduction of the dynamical equations can be made. Moreover, from bulk considerations it can be seen that the presence of the Dresselhaus term modifies the band structure and, consequently, it influences the magnitude of the energy mismatch responsible of the confinement.

It is worth to mention that the proposed dots rely on materials characterized by bulk spin-orbit intensities of the order of 10^{-9} eVcm and a range of tunability of the same order in order to generate the required spin-orbit gradient. These values correspond to heterostructures based on narrow gap semiconductors; in this sense InGaAs parameters could be enough to observe this type of confinement, but InAs-based heterostructures reveals as the best candidates even to observe collective states. A relevant feature of this materials is that they show a relevant coupling between the different semiconductor bands, therefore, the conduction-band effective mass Hamiltonian is quite limited in order to model them. In this sense we would like to remark that the numerical results reported in this chapter must be taken in a qualitative way and that a multiband treatment should be necessary in order to obtain a more quantitative results.

References

[1] L. Jacak, P. Hawrylak, A. Wójs. *Quantum Dots*. Ed. Springer (1998) p. 176.

[2] V. Fleurov, V. A. Ivanov, F. M. Peeters, I. D. Vagner, *Physica E* **14**, 361-365 (2002).

[3] Yu. V. Pershin (*cond-mat/0305002*).

[4] G. Dresselhaus, *Phys. Rev.* **100**, 580 (1955).

[5] E. I. Rashba, Fiz. tverd. Tela (Leningrad) **2**,1224 (1960) [*Sov. Phys. Solid State* **2**, 1109 (1960)].

[6] J. Nitta, T. Akazaki, H. Takayanagi and T. Enoki, *Phys. Rev. Lett.* **78**, 1335 (1997).

[7] C.-M. Hu, J. Nitta, T. Akazaki, H. Takayanagi, J. Osaka, P. Pfeffer and W. Zawadzki, *Phys. Rev. B* **60**, 7736 (1999).

[8] D. Grundler, *Phys. Rev. Lett.* **84**, 6074 (2000).

[9] E. Ya. Sherman, *Phys. Rev. B* **67**, 161303 (2003)

[10] E. Ya. Sherman, *Appl. Phys. Lett.* **82**, 209 (2003)

[11] P. W. Brouwer, J. N. H. J. Cremers, B. I. Halperin, *Phys. Rev. B* **65**, 081302 (2002).

[12] M. Valín-Rodríguez, A. Puente, Ll. Serra, *Nanotech.* **14**, 882-885 (2003).

[13] E. A. de Andrada e Silva, G. C. La Rocca, F. Bassani, *Phys. Rev. B* **55**, 16293 (1997).

[14] T. Darnhofer, U. Rössler, *Phys. Rev. B* **47**, 16020 (1993).

[15] W. Knap, C. Skierbiszewski, A. Zduniak, E. Litwin–Staszewska, D. Bertho, F. Kobbi, J. L. Robert, G. E. Pikus, F. G. Pikus, S. V. Iordanskii, V. Mosser, K. Zekentes, and Yu. B. Lyanda–Geller, *Phys. Rev. B* **53**, 3912-3924 (1996).

[16] I.D. Vagner, A.S. Rozhavsky, P. Wyder, A. Yu. Zyuzin, *Phys. Rev. Lett.* **80**, 2417 (1998).

[17] A.G. Aronov, Y.B. Lyanda-Geller, *Phys. Rev. Lett.* **70**, 343 (1993).

[18] M. Valín-Rodríguez, A. Puente, Ll. Serra, *Phys. Rev. B* **66**, 045317 (2002).

In: Frontiers in Quantum Dots Research
Editor: Marco L. Voronin, pp. 69-83
ISBN: 1-60021-057-0

Chapter 3

WIGNER CRYSTALLIZATION IN A FEW-ELECTRON CIRCULAR QUANTUM DOT

Nurten Akman[1] and Mehmet Tomak[2]
[1]Department of Physics, Mersin University, 33342 Mersin, Turkey
[2]Department of Physics, Middle East Technical University,06531 Ankara, Turkey

Abstract

The charge density and pair correlation function of three interacting electrons confined in a two-dimensional circular quantum dot are calculated by the exact diagonalization of the Hamiltonian. The circular quantum dot is characterized by the hard wall type confinement potential to obtain the observable effects of the electron-electron interactions on the electron distribution within the quantum dot. For the circular quantum dot geometry, in addition to the total spin S, S_z and the z- component of the total orbital angular momentum M are good quantum numbers. Furthermore, both single particle energy level and matrix element of the Coulomb potential are independent of the spin and sign of m. All of these symmetry properties cause us to interest in many electron states having fixed values of M and S_z for a given electron number, N. Also we restrict ourselves to those states for which $S_z = S$ to work out the relevant quantities in detail. In the analytical and numerical calculations of the charge density and pair correlation function, we assume spin polarized three electrons with vanishing total orbital angular momentum, $M = 0$. Variation of the charge density with radial distance as well as that of the pair correlation function with azimuthal angle together with their size dependence show clearly the atomic limit and the formation of the Wigner molecule. The Wigner-molecular arrangement of three electrons occurs in the form of an equilateral triangular structure as the size of the dot is increased. In this quasi-classical configuration, the electrons minimize their electrostatic energy only by arranging themselves at the vertices of an equilateral triangle.

PACS Numbers: 71.10.Hf, 73.21.-b

1. Introduction

Adventure of the low dimensional systems began in the early 70's, when the first two dimensional quantum wells were successfully manufactured by the thin-film epitaxial techniques. It is now possible to grow ultrahigh purity thin films of compound semiconductors by the well controlled epitaxial growth techniques like MBE and MOCVD. Each layer can be precisely controlled down to a few Å thickness. Current nanotechnology have allowed the lateral confinement of a two dimensional electron gas (2DEG) by means of suitably shaped gate electrodes or by etching techniques [1-4]. Various theoretical and experimental research have proven that when the system size approaches the electronic wavelength, quantum mechanical effects become important and determine the properties of the system. Typical dimensions of these confined systems range from nanometers to a few microns.

Confined systems having a discrete energy spectra are commonly called as "zero-dimensional systems" or "quantum dots" [5-16]. The motion of electrons in a quantum dot is quantized in all three spatial directions, all having dimensions comparable to the de Broglie wavelength of the particle $(L_x < L_y < L_z < \lambda_F)$. However, if the quantization in the vertical direction is much stronger than the quantization in the in-plane directions, a quantum dot could be treated as disc-like 2D system where electrons have significant freedom along x- and y- directions.

Experimentally, the size and shape of a quantum dot [17] and the number of electrons confined in it [18-21] could be altered simply by changing the gate voltage applied. The number of electrons in these systems might be as few as 2, 3, or 4 like in natural atoms or in the order of millions. If the number of electrons is large, the electronic properties of dots may be qualitatively described by the well-known properties of the infinite electron gas. For a few number of electrons, however, their electronic properties have a strong particle number dependence [22]. Few electron quantum dots have already been realized and investigated by optical absorption experiments [11-13,16]. Since many electronic properties of dots resembles those of atoms such as discrete energy spectrum and comparable number of electrons, they are also regarded as "artificial atoms". In a recent review [23], the current-voltage measurements are used to observe this resemblance. In addition, varying the number of electrons on the dot in a controlled way allows one to scan the entire periodic table of artificial atoms by simply changing the gate voltage.

The Coulomb interaction in these artificial atoms is very important for understanding their quantum mechanical properties. Especially correlations among electrons are crucial since their effects influence the spectral [13-16,24] and transport properties [25-27] of quantum dots. So far, various analytical techniques have been devised for handling the electron correlations in quantum dot systems. One of these techniques include the solution of the many particle Schrödinger equation [28]. In this three dimensional approach, interacting electrons in non-parabolic quantum dot systems have been investigated using the formalism of Hylleraas [29] where interelectron coordinates are built into the wavefunction explicitly. By comparing Hylleraas and Hartree-Fock type results, in [28], it is shown how the charge redistributions, attributable to correlation interactions, affect ground and excited state energies and electron confinement in isolated and coupled two quantum dot systems. One- and two-electron ground state energies of a silicon sphere embedded in an amorphous silicon dioxide matrix [30] have been calculated as a function of sphere size. The electron-electron interaction and polarization effects in that study have been treated by perturbation.

A great deal of interest has also gone into analytical investigation of correlation effects in 2D quantum dot systems. Comparison of energies, pair correlation functions, and particle densities of the singlet and triplet ground state of quantum dot helium in a magnetic field has been investigated in [24] by using Hartree, Hartree-Fock, and exact diagonalization methods. The HF results for the triplet ground state were found to be in very good agreement with the exact diagonalization results, proving the importance of the exchange interaction. On the other hand, the exact results for the singlet state have disagreements with the HF results. In [24] it has been shown that the disagreement between two results arises due to the electron correlations which are neglected in the Hartree-Fock approximation. However, most of the work performed on the electron-electron correlations in quantum dots placed in a magnetic field [9,11,31], as well as transport experiments [32] and far-infrared spectroscopy [33] has been based on a two dimensional quantum dot with parabolic type confinement potential.

Commonly the lateral confinement potential in quantum dots is created by electrostatically squeezing the planar charge distribution in the 2DEG. Therefore, it shows approximately a parabolic behaviour ($\sim r^2$, instead of the $1/r$ attractive potential of the atoms) and represents reasonably well the electrostatically confined electrons. The advantage of using a harmonic type of confining potential is the analytical simplicity of the problem since the center of mass and relative coordinates decouple. However, in the far-infrared spectroscopy, the radiation field can not detect the electron-electron correlations when the confinement is parabolic. This is caused by the decoupling of far-infrared radiation with the relative motion of electrons [34].

Thus, in order to probe the interaction effects, it has been suggested that the shape of the dot should be modified to achieve the coupling of center of mass and relative motions [34]. In [34], the heat capacity results have also been presented to suggest that the interaction effects could be seen by measuring the thermodynamic properties of the electrons. The magnetization of parabolic quantum dots has been also computed and magnetization found to be another probe of the interaction effects [35] like the heat capacity. The magnetization of the dots is predicted to oscillate with magnetic field because the ground state prefers to be at certain magic values of the total angular momentum. This behaviour is explained in a subsequent work [36] as a direct consequence of the Pauli principle which enables the electrons to reduce their energy optimally only at the magic angular momenta. Recently, vertically coupled quantum dots or artificial molecules [37,38] have attracted considerable attention. In [38], a double dot system with three spin-polarized electrons was investigated and a sequence of angular-momentum magic numbers was found depending on the strength of the interdot tunneling. Besides these, some authors have achieved the coupling of center of mass and relative motions of electrons with deviations from the exact harmonic confinement [39-41]. With etching techniques or self organized growth it is possible to create hard-wall type confinement potential [42]. Since the center of mass motion of electrons is coupled with their relative part in this type of confinement potential, the effects of correlation influencing energy spectrum, transport, and spectral measurements can be studied with the excitation spectra [43].

In the present article we analyze electron correlation effects in a 2D circular quantum dot with hard wall confinement potential. We utilize exact diagonalization technique in computing the eigenvalues and corresponding eigenvectors for the ground state numerically.

We specialize to the case of three spin-polarized electrons with total magnetic quantum number S_z=3/2 and total orbital angular momentum M=0. We investigate charge density and pair correlation function as a function of the dot size under these circumstances. The observation of Wigner molecule, in equilateral triangular configuration of three electrons, as the dot size increases is discussed qualitatively. In Sec. 2 we give the formalism and details of the calculation. In Secs. 3 and 4 we report and discuss the results of the numerical calculations.

2. Formalism

We consider few electron system in a two dimensional circular quantum dot of radius a. The longitudinal and the transverse coordinates are denoted as z and $\vec{r}=(\rho,\theta)$ respectively, so that the three--dimensional coordinate is $\vec{x}=(z,\vec{r})$. The width of the quantum dot is assumed to be much smaller than its radius, $w << a$, in the analytical and numerical calculations.

An N- electron system is described by the Hamiltonian

$$H=\sum_{i=1}^{N}\{-\frac{\hbar^2}{2m^*}\vec{\nabla}_i^2+U(\vec{x}_i)\}+\frac{e^2}{2\varepsilon}\sum_{i\neq j}\frac{1}{|\vec{x}_i-\vec{x}_j|}, \quad (1)$$

where m^* is the electron effective mass and ε is the dielectric constant of the medium. The second term in the parenthesis, $U(\vec{x})$ is the hard-wall type confinement potential which vanishes for $|\vec{x}|<a$, and infinite for $|\vec{x}|\geq a$. By using a as the unit of length and by inserting the dimensionless coupling constant $\lambda = a/a_B$ and the effective Bohr radius $a_B=\frac{\varepsilon\hbar^2}{m^*e^2}$ into the equation above, we obtain the dimensionless Hamiltonian $\mathcal{H}$ via the definition

$$H=\frac{\hbar^2}{m^* a_B^2\lambda^2}\mathcal{H}, \quad (2)$$

where

$$\mathcal{H}=\sum_{i=1}^{N}(-\frac{1}{2})\vec{\nabla}_i^2+\frac{\lambda}{2}\sum_{i\neq j}\frac{1}{|\vec{x}_i-\vec{x}_j|}. \quad (3)$$

One recalls that the energy is measured in units of $\hbar^2/(m^* a_B^2\lambda^2)$ which depends on λ, and the length is measured in units of a, the dot radius. In (3), the Coulomb potential is multiplied by $\lambda = a/a_B$ characterizing the strength of the interaction. Although it is possible to perform a perturbative expansion in powers of λ to calculate the energy spectrum [44], when λ exceeds unity perturbation theory does not work and one has to resort to an applicable method that covers a wide range of λ. Renormalized perturbation theory is applicable at the values of $\lambda > 1$ [45] but we prefer to use the exact diagonalization technique to get the ground state energies and eigenvectors as in Refs. [43,46].

To express the dimensionless Hamiltonian $\mathcal{H}$ in the second quantization language, one inserts the density operator $\rho(\vec{x}) = \Psi^{\dagger}(\vec{x})\Psi(\vec{x})$ into Eq.(3). Here $\Psi(\vec{x}) = \sum_K a_K \varphi_K(\vec{x})$ is the many particle eigenstate with annihilation operator a_K and single particle eigenstate $\varphi_K(\vec{x})$. After some steps, the new $\mathcal{H}$ is described by

$$\mathcal{H} = \sum_K \mathsf{E}_K a_K^{\dagger} a_K + \frac{\lambda}{2} \sum_{K,L,M,N} V_{K,L,M,N} a_K^{\dagger} a_L^{\dagger} a_M a_N, \tag{4}$$

where

$$\mathsf{E}_K = \int d^2x \varphi_K^{\dagger}(\vec{x})(-\frac{1}{2})\vec{\nabla}^2 \varphi_K(\vec{x}), \tag{5}$$

and

$$V_{K,L,M,N} = \int\int d^2x d^2x' \mathrm{V}(\vec{\mathrm{x}} - \vec{\mathrm{x}}')\varphi_K^{\dagger}(\vec{x})\varphi_L^{\dagger}(\vec{x}')\varphi_M(\vec{x}')\varphi_N(\vec{x}). \tag{6}$$

In these equations $\varphi_A(\vec{x})$ is again the single particle eigenstate of the free Hamiltonian, and $A(=K,L,M,N)$ is a collective index designating the radial (n_A), angular momentum (m_A) and spin (σ_A) quantum numbers. From the free Hamiltonian one can easily get

$$\varphi_A(\vec{x}) = \phi_{n_A,m_A}(\vec{x})\chi_{\sigma_A}, \tag{7}$$

where χ_σ is the spin wavefunction, and $\phi_{n,m}(\vec{x})$ is the normalized orbital eigenfunction

$$\phi_{n,m}(\vec{x}) = \frac{1}{\sqrt{\pi}} \frac{1}{\left|J_{|m|+1}(k_{n,|m|})\right|} e^{im\theta} J_{|m|}(k_{n,|m|}|\vec{x}|). \tag{8}$$

Using Eqs. (7) and (8) in (5) one gets the eigenvalues $\mathsf{E}_K = (1/2)k^2_{n_K,|m_K|}$ which are independent both of the spin and sign of m. It depends only on the n_K - th root of the Bessel function of the first kind J_{m_K}. Here $k_{n,|m|}$ are the zeroes of the Bessel function $(J_{|m|}(k_{n,|m|}) = 0)$.

Since the one-electron orbitals in Eq. (8) depend already on the Bessel functions, it is practically convenient to expand the Coulomb potential, $V(\vec{x} - \vec{x}') = 1/|\vec{x} - \vec{x}'|$, in terms of the Bessel functions as well

$$\frac{1}{|\vec{x} - \vec{x}'|} = \sum_{m=-\infty}^{\infty} \int_0^{\infty} dk\, e^{im(\theta-\theta')} J_{|m|}(k\rho) J_{|m|}(k\rho') e^{-k(z_> - z_<)}, \tag{9}$$

where $(z_> - z_<)$ shows the extension of the dot in the longitudinal direction. If the longitudinal extension of the dot is a non-negligible fraction of its in-plane size, one has to analyze the excitations in this direction, too. However, in the limit of $(z_> - z_<) \to 0$, it is sufficient to probe only the transverse plane assuming that the system is in the lowest state for longitudinal dynamics. In the following, we follow the latter one and assign a small value to $(z_> - z_<)$ in the calculations. In this approximation, by using Eqs. (7), (8) and (9) in (6), the matrix element of the Coulomb potential becomes

$$\begin{aligned} V^{n_K,n_L,n_M,n_N}_{m_K,m_L,m_M,m_N} &= 4 \frac{1}{\left|J_{|m_K|+1}(k_{n_K,|m_K|})\right|} \frac{1}{\left|J_{|m_L|+1}(k_{n_L,|m_L|})\right|} \\ &\frac{1}{\left|J_{|m_M|+1}(k_{n_M,|m_M|})\right|} \frac{1}{\left|J_{|m_N|+1}(k_{n_N,|m_N|})\right|} \int_0^{\infty} dk \int_0^1 d\rho\, \rho \\ &J_{|m_K|}(k_{n_K,|m_K|}\rho) J_{|m_L|}(k_{n_L,|m_L|}\rho) J_{|m_K - m_L|}(k\rho) \\ &\int_0^1 d\rho'\, \rho' J_{|m_M|}(k_{n_M,|m_M|}\rho') J_{|m_N|}(k_{n_N,|m_N|}\rho') J_{|m_N - m_M|}(k\rho'). \end{aligned} \tag{10}$$

One notes that the angular integrations in $V_{K,L,M,N}$ impose the selection rule $m_K - m_L = m_N - m_M$, otherwise $V_{K,L,M,N}$ vanishes.

The total orbital angular momentum $M = \sum_{i=1}^{N} m_i$, the total spin S and the total magnetic quantum number $S_z = \sum_{i=1}^{N} s_i$ are good quantum numbers for N electrons in the dot due to the circular geometry of the dot, and spin independence of the Coulombic interaction. For the sake of these symmetry properties we will be interested in those many electron states having fixed values of M and S_z for a given electron number, N. Moreover, we will restrict ourselves to those states for which $S_z = S$.

For calculating the physical quantities such as energy spectrum, charge density and pair correlation function, one needs to compute the matrix elements of the Hamiltonian operator between N-electron ground state which can be expressed as

$$\left|\Phi_0^{(N)}\right\rangle = \sum_C b_C^0 \left|C\right\rangle, \tag{11}$$

where $\left|C\right\rangle = a^{\dagger}_{n_1,m_1,\sigma_1} \dots a^{\dagger}_{n_N,m_N,\sigma_N} \left|0\right\rangle$ is a non-interacting Slater determinant and sum runs over all possible configurations of the quantum numbers (n, m, σ) satisfying given values of M and S_z. The expansion coefficients b_C^0 are identified with the eigenvectors of the ground state. In this work we are dealing with the quartet ground state, $\left|S_z = 3/2, M = 0\right\rangle$,

$$\left|S_z = 3/2, M = 0\right\rangle = a^{\dagger}_{n_1,m_1,\uparrow} a^{\dagger}_{n_2,m_2,\uparrow} a^{\dagger}_{n_3,m_3,\uparrow} \left|0\right\rangle \tag{12}$$

in which all electrons are spin polarized. We will compute the relevant quantities in the quartet ground state. With this basis state in hand, one can construct the Hamiltonian matrix by evaluating the matrix elements of (4). The Hamiltonian matrix is formed in the space of (n_1, m_1), (n_2, m_2) and (n_3, m_3) under the constraints related with the fixed M and S_Z values. Due to the large amount of computer memory needed for the evaluation of the Hamiltonian matrix elements, the eigensolution of the Hamiltonian could be obtained only for very small N. Also it is necessary to cut off n and m at a large enough value for not to louse up the precision of the computation. It is clear that this procedure truncates the semi-infinite Hamiltonian matrix to a finite one. Diagonalization of the later yields the energy eigenvalues and the eigenvectors, b_C^0 of the few electron system. After obtaining the expansion coefficients b_C^0 in this way, one turns to the computation of other physical quantities such as charge density and the pair correlation function.

The charge density of the electrons are defined by

$$\rho(\vec{x}) = \sum_{\sigma} \left\langle \Phi_0^{(N)} \left| \phi_\sigma^{\dagger}(\vec{x}) \phi_\sigma(\vec{x}) \right| \Phi_0^{(N)} \right\rangle, \tag{13}$$

which measures the electron density at a given point $\vec{x}$ in space. A close inspection of this formula reveals that, at a fixed radius, $\rho(\vec{x}) \sim \cos(2M\theta)$ so that angular dependence of the charge density is determined by the total orbital angular momentum. Hence, for $M = 0$ angular dependence disappears and $\rho(\vec{x})$ assumes only a radial variation dictated by the associated Bessel functions.

Another relevant quantity, the pair correlation function, is a two-point function defined by

$$\rho_c(\vec{x},\vec{x}') = \sum_{\sigma,\sigma'} \left\langle \Phi_0^{(N)} \middle| \phi_\sigma^\dagger(\vec{x})\phi_{\sigma'}^\dagger(\vec{x}')\phi_{\sigma'}(\vec{x}')\phi_\sigma(\vec{x}) \middle| \Phi_0^{(N)} \right\rangle, \tag{14}$$

which is the probability of finding an electron at $\vec{x}$ given that another one is situated at $\vec{x}'$.

In the next section we will compute the charge density and pair correlation function in order to identify their dependence on the space coordinates as well as the dot size λ.

3. Numerical Analysis

Electron-electron correlations have been computed for a 1D hard-wall type dot in [47], and a 2D parabolic dot in [48]. In the latter the Schroedinger equation is solved for three electrons numerically for a fixed dot size. For three electrons with $S_z = 3/2$ and $M = 3k$ $(k = 0,1,2, ...)$, it is expected that, in the ground state, electrons form an equilateral triangle at some radius r_0 determined by the confining parabolic potential. Here we analyze charge density and pair correlation function for three electrons in a circular quantum dot for $S_z = 3/2, M = 0$. In contrast to the parabolic confinement potential, in the case of hard-wall type confinement one cannot find an analitic expression for r_0. Despite of this, however, for large enough λ value, the equilateral triangular configuration is expected to occur at some distance in the radial direction. This then will be an indication of the Wigner molecule structure which is a general feature of quantum dots for large λ [47]. This feature is also observed in [49] where the behaviour of the ground state charge density for N=3 and N=4 is in a qualitative agreement with that of Ref. [47].

Fig.1 describes the r and λ dependences of the charge density $\rho(\vec{x})$ (normalized to N=3 for each λ) for the state $|S_z = 3/2, M = 0\rangle$. Three curves shown in this figure stand for the variations of the electron distribution as a function of λ and r. For small λ (e. g. $\lambda = 1$) electrons have non-negligible distribution at the center of the dot though maximum is reached away from the origin. This behaviour of $\rho(\vec{x})$ proves that for small λ, $J_{m=0}(r)$ is the dominant component especially at small r. As λ starts to increase, however, the charge density vanishes gradually around the center of the dot (e. g. $\lambda = 10, 100$). It here that one observes the dominance of $J_{m>0}$ since they vanish at origin by definition.

Besides the behaviour of the charge density close to the center of the dot, one also observes that the higher the λ, the sharper the peaks. The sharper peaks show the most probable radial position of the electrons. Additionally, as λ increases, the electrons get shifted towards the edge of the dot, but they never reach there because of the infinite potential barrier. For example, for $\lambda = 100$ the electrons are distributed most probably along the periphery of the circle whose radius is $r \sim 0.7$ within the quantum dot. This result follows from the fact that the system minimizes its electrostatic energy which is the dominant part of the total Hamiltonian for large enough λ values. The behavior of the charge density informs us only about the radial distribution of electrons. Therefore, to have a complete knowledge of the planar electron-electron correlations, we must also investigate the variation of the pair correlation function $\rho_c(\vec{x},\vec{x}')$ with r, θ as well as λ.

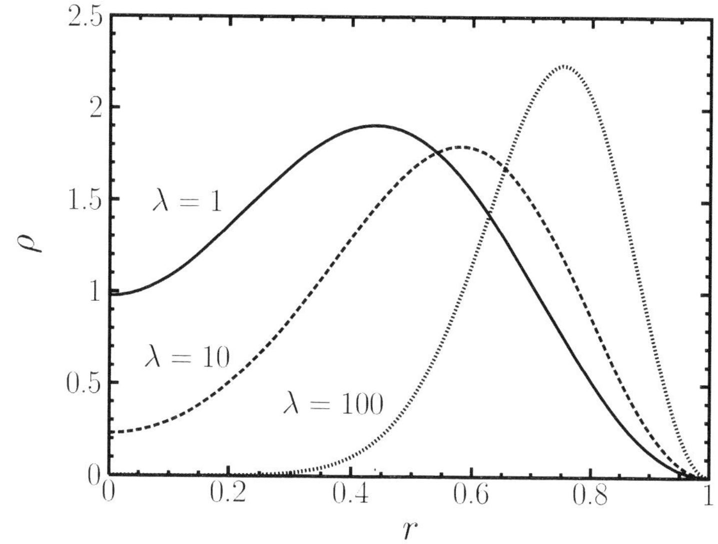

Figure 1. Variation of the charge density $\rho(\vec{x})$ with radial distance r and the dot size λ.

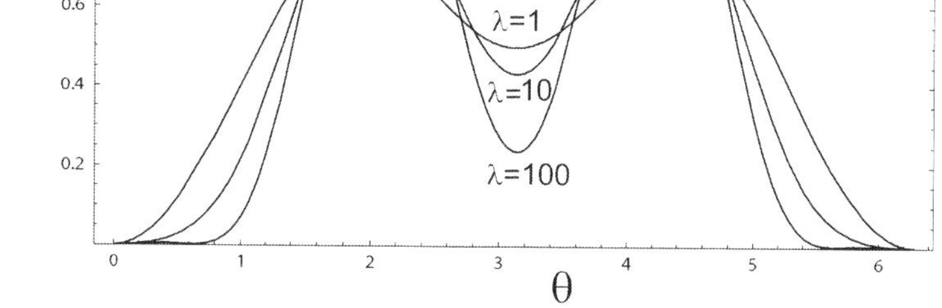

Figure 2. Variation of the pair correlation function with θ and λ for $r = r' = 0.67$ and $\theta' = 0$.

In Fig. 2 the variation of the pair correlation function (normalized to N(N-1)=6) with θ and λ for $r = r' = 0.67$ and $\theta' = 0$ is presented. The pair correlation function $\rho_c(\vec{x},\vec{x}')$ is the probability distribution of N-1 electrons when one of the N electrons is located at $\vec{x}'$. In the case of three electrons, the pair correlation function, in Fig. 2, shows the angular distribution of two electrons within the circular quantum dot. The fact that the pair correlation function has always two distinct peaks and vanishes at $\theta = 0$, as is seen in this figure, the Pauli exclusion principle is satisfied as expected.

An important property of the pair correlation function follows from its variation with the dot size in units of the effective Bohr radius. One notices that for small λ values, peaks are not sharp and electrons do not have a well-defined configuration. Namely, the pair correlation function is not diminished significantly between the two peaks. The small λ limit shows nothing but the atomic regime of confinement where dot size is of the order of effective Bohr radius or smaller; and their average kinetic energy exceeds the Coulombic repulsion. In this limit, it is well known that the electrons are completely delocalized in the quantum dot. That is, they do not possess well-defined positions in the plane of the dot. It is also known that the perturbation theory is a reliable tool to investigate the physical properties of the dot [44] for the values of $\lambda < 1$. Indeed, the Hartree-Fock method can be applied to the dot structures just as in the few-electron natural atoms [50]. Furthermore, Fig. 2 shows clearly that the peaks get sharper with increasing λ and the pair correlation function assumes gradually two distinct peaks between which there is a strong depletion. Therefore, the higher the dot size, the smaller the overlap between the two peaks. Indeed, as λ increases the peaks approach to fixed positions such that their angular seperation is $\sim 120^{\circ}$. This angular seperation constitutes an equilateral triangular structure as already emphasized in other investigations [48,50]. This geometrical rearrangement corresponds to the minimum energy configuration of the system where its kinetic energy is smaller than the Coulomb energy. The latter is minimized by the equilateral triangular configuration of the electrons. The approach to this configuration for large dot sizes is nothing but the formation of the well-known Wigner molecule structure which has been shown to exist in other types of dots too [47,50]. As a result, for large dot sizes the electrons locate at fixed positions in the dot by minimizing their electrostatic energy.

For a better understanding of the behavior of the electron distribution in the dot, it may be convenient to analyze the pair correlation function in the $r - \theta$ plane in which both radial and angular variations become visible. Fig. 3 shows the contour lines of $\rho_c(\vec{x}, \vec{x}')$ in $r - \theta$ plane for $\theta' = 0$, $r' = 0.67$, and $\lambda = 1$. In accordance with Fig. 1 and Fig. 2, there is a non-negligible correlation between the electrons for small r; also the maxima are reached away from the origin, $r \sim 0.4$. Moreover, near the edge of the dot, the correlation distribution is highly suppressed. One further notices that the pair correlation function in between the two peaks is not suppressed at all.

To illustrate the effects of larger dot sizes on Fig. 3, the pair correlation function in the $r - \theta$ plane is shown in Fig. 4 for $\lambda = 100$. As is seen, there are important differences between Fig. 3 and Fig. 4. First of all, the peaks are shifted towards the edge of the dot in comparison with Fig. 1. Next the peaks are now sharpened and the pair correlation between them is reduced significantly. In this sense, the differences between these two figures prove the formation of the Wigner molecule of the electrons with increasing Coulomb repulsion among them with ever growing λ.

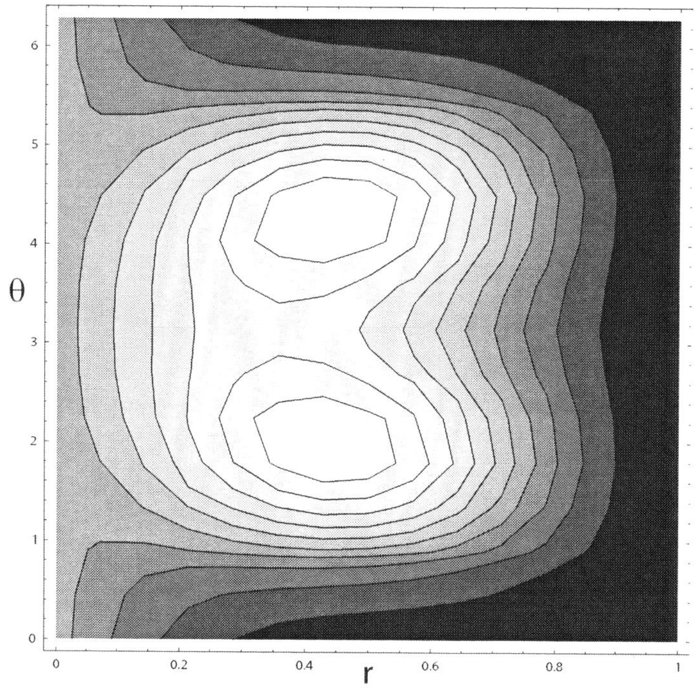

Figure 3. Contour lines showing the pair correlation function on $(r-\theta)$ plane for $r'=0.67, \theta'=0$ and $\lambda=1$.

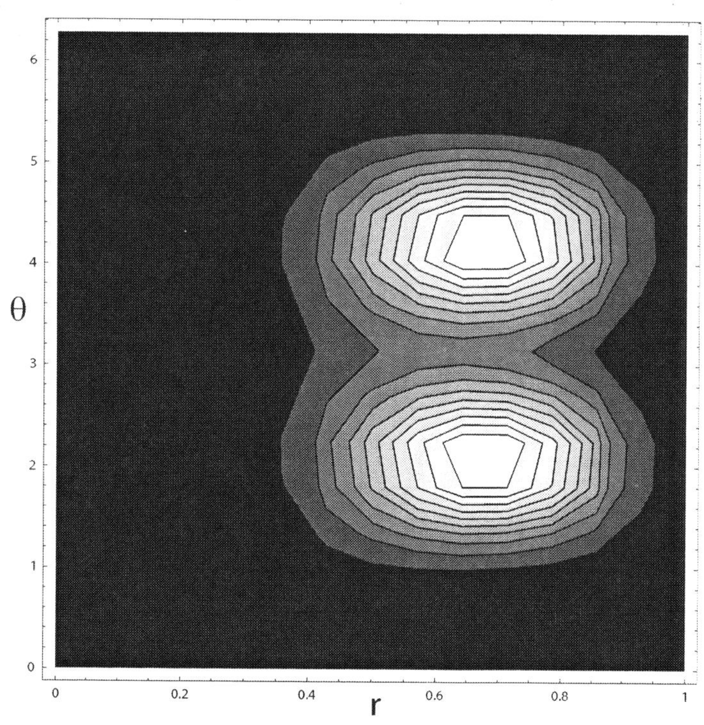

Figure 4. The same as in Fig. 3 but for $\lambda=100$.

4. Conclusion

Recently, three-electron system confined to a hard spherical box of radius R was studied by the use of unrestricted Hartree-Fock and full configuration-interaction methods for high, medium, and low density regimes [51]. They observed that the three-electron system changes its configuration from s^2p to sp^2 in the large R limit. The latter configuration corresponds to an equilateral triangle which is precisely the shape of the Wigner molecule required to minimize the Coulomb repulsion between the three electrons.

In [52], a detailed analysis of unrestricted Hartree-Fock method for up to eigth electrons in a parabolic quantum dot provides rather accurate estimates of the ground state energy. In the weak interaction limit shell model characteristics and in the strong interaction Wigner molecule feature were observed for the number of particles under consideration.

By using the unrestricted variational Hartree-Fock method for $T = 0$ and quantum path integral Monte Carlo method for $T \neq 0$, the behaviour of three electrons inside a spherical quantum dot with parabolic confinement was studied in detail [53]. Quantum transitions which involve a spin transformation and a phase transition from a Wigner crystal like state to a Fermi liquid like one obtained as the dimensionless quantum control parameter q is increased. Since q gives the dimensionless ratio of the characteristic quantum kinetic energy to the characteristic quantum Coulomb interaction, for small q, the spin configuration of the ground state of the three electron system is ferromagnetic with total spin $S = 3/2$. Besides the maxima of the electron density are located at the corners of an equilateral triangle for small q. At a critical q value it is seen that the ground state spin changes from $S = 3/2$ to $S = 1/2$ and also asymmetry appears in the triangle. At a second critical value of q, the electron distribution experiences a phase transition from trianglelike to axial symmetric and as q increases further, a Fermi liquid like electron distribution in the ground state with $S = 1/2$ is obtained.

Very recently a growing interest has increased in obtaining the crossover from Fermi liquid to Wigner crystal in quantum dots. R. Egger *et al.* [54] extracted the crossover value $r_c \sim 4$ from the path integral Monte Carlo simulations for up to eigth electrons confined in a parabolic quantum dot. Their numerical results are in aggreement with the Wigner molecule description for $r_s > r_c$ and Fermi liquid like distribution is clearly seen for $r_s < r_c$. The obtained r_c is very small when compared to the 2D electron gas [55,56] in which $r_c \sim 37$ for the homogoneous system and $r_c \sim 7.5$ for the disordered system.

In [57], the energy spectra, charge density, pair correlation function and some thermodynamical properties of quantum dot lithium were studied with the method of exact diagonalization similar to our previous work [58]. Full energy spectra in [57] yields a level crossing in the ground state of the dot at the critical value of λ $(\lambda = \lambda_c = 4.343)$. The total spin of the system in the ground state changes at this critical point from $S_{tot} = 1/2$ at $\lambda < \lambda_c$ to $S_{tot} = 3/2$ at $\lambda > \lambda_c$. The electron density in the state $(0, 3/2)$, where $L_{tot} = 0$ and $S_{tot} = 3/2$, does not depend on the angular coordinate θ due to the rotational symmetry of the Hamiltonian, and so its variance as a function of r describes the distribution of three electrons within the quantum dot quite well. The electrons have a non-negligible distribution around the origin of the dot for small λ values but with the growing interaction parameter λ, the minimum of the electron density at $r = 0$ is reduced and the maxima shift to the right which means that the electrons go away from origin of the dot. In the pair correlation function plot, the overlapping between the two peaks is clearly seen for small λ values where the electron-electron interaction is weak. As λ increases the overlapping between the peaks decreases and for a strong Coulomb interaction $(\lambda \geq 4)$, the Wigner molecule in the shape of the equilateral triangle arises. All these results presented in [57] resemble to our charge density and pair correlation function obtained for the same number of electrons and the ground state.

In this work we have performed a detailed numerical study of the electron correlation effects in a 2D circular quantum dot with hard wall confinement potential. Our investigations show that as the dot size increase gradually configuration of the electrons approach an equilateral triangular structure. Hard-wall quantum dots, which may be as interesting as parabolic ones for experimental studies, allow for the formation of the Wigner molecule structure of the electrons for large dot sizes compared to the Bohr radius. It is known that the optical properties like the inelastic light scattering or far infrared absorption [59] and magnetic properties are all dependent on the spin of the ground state. Therefore, experimental studies on the quantum dots can reveal valuable information on the electron configuration in the dot by concentrating on the spin dependent quantities.

References

[1] J. N. Randall, M. A. Reed, T. M. Moore, R. J. Matyi, and J. W. Lee, *J. Vac. Sci. and Tech.* **B 6** (1988) 302.

[2] M. A. Reed, J. N. Randall, R. J. Aggarwal, R. J. Matyi, T. M. Moore, and A. E. Wetsel, *Phys. Rev. Lett.* **60** (1988) 535.

[3] U. Meirav, P. L. McEuen, M. A. Kastner, E. B. Foxman, A. Kumar, and S. J. Wind, *Z. Phys.* **B 85** (1991) 357.

[4] G. W. Bryant, *Phys. Rev. Lett.* **59** (1987) 1140.

[5] M. A. Reed, J. N. Randall, J. Aggarwal, R. J. Matyi, T. M. Moore, and A. E. Wetsel, *Phys. Rev. Lett.* **60** (1988) 535.

[6] W. Hansen, T. P. Smith III, K. Y. Lee, J. A. Brum, C. M. Knoedler, J. M. Hong, and D. P. Kern, *Phys. Rev. Lett.* **62** (1989) 2168; 64 (1990) 1991; W. Hansen, T. P. Smith III, K. Y. Lee, J. M. Hong, and C. M. Knoedler, *Appl. Phys. Lett.* **56** (1990) 168.

[7] L. P. Kouwenhoven, F. W. J. Hekking, B. J. van Wees, C. J. P. M. Harmans, C. E. Timmering, and C. T. Foxon, *Phys. Rev. Lett.* **65** (1990) 361.

[8] P. L. McEuen, E. B. Foxman, U. Meirav, M. A. Kastner, Y. Meir, N. S. Wingreen, and S. J. Wind, *Phys. Rev. Lett.* **66** (1991) 1926.

[9] Ch. Sikorski and U. Merkt, *Phys. Rev. Lett.* **62** (1989) 2164; 64 (1990) 3100; *Surf. Sci.* **229** (1990) 282.

[10] Lorke, J. P. Kotthaus, and K. Ploog, *Phys. Rev. Lett.* **64** (1990) 2559.

[11] T. Demel, D. Heitmann, P. Grambow, and K. Ploog, *Phys. Rev. Lett.* **64** (1990) 788.

[12] Meurer, D. Heitmann, and K. Ploog, *Phys. Rev. Lett.* **68** (1992) 1371.

[13] W. Hausler and B. Kramer, *Phys. Rev.* **B 47** (1993) 16353.
[14] P. Hawrylak and D. Pfannkuche, *Phys. Rev. Lett.* **70** (1993) 485.
[15] U. Merkt, J. Hausler, and M. Wagner, *Phys. Rev.* **B 43** (1991) 7320.
[16] W. Hausler, B. Kramer, and J. Masek, *Z. Phys.* **B 85** (1991) 435.
[17] G. W. Bryant, *Phys. Rev. Lett.* **59** (1987) 1140.
[18] Ch. Sikorski and U. Merkt, *Phys. Rev. Lett.* **62** (1989) 2164.
[19] Ch. Sikorski and U. Merkt, *Phys. Rev. Lett.* **64** (1990) 3100.
[20] Ch. Sikorski and U. Merkt, *Surf. Sci.* **229** (1990) 282.
[21] Meurer, D. Heitmann and K. Ploog, *Phys. Rev. Lett.* **68** (1992) 1371.
[22] N. F. Johnson, *J. Phys.: Condens. Matter* **7** (1995) 965.
[23] L. P. Kouwenhoven, D. G. Austing and S. Tarucha, *Rep. Prog. Phys.* **64** (2001) 701.
[24] Pfannkuche, V. Gudmudsson, and P. A. Maksym, *Phys. Rev.* **B 47** (1993) 2244.
[25] Y. Meir, N. S. Wingreen, and P. A. Lee, *Phys. Rev. Lett.* **66** (1991) 3048; J. M. Kinaret, Y. Meir, N. S. Wingreen, P. A. Lee, and X. G. Wen, *Phys. Rev.* **B 46** (1992) 4681.
[26] J. J. Palacios, L. Martin-Moreno, and C. Tejedor, *Europhys. Lett.* **23** (1993) 495.
[27] S. R. Patel, D. R. Stewart, C. M. Marcus, M. Gokcedag, Y. Alhassid, A. D. Stone, C. I. Duruoz, and J. S. Harris, Jr, *Phys. Rev. Lett.* **81** (1998) 5900.
[28] H. Guerrero, *Semi. Sci. Tech.* **10** (1995) 759.
[29] A. Hylleraas, *Z. Phys.* **54** (1929) 347.
[30] D. Babic, R. Tsu, and R. F. Greene, *Phys. Rev.* **B 45** (1992) 14150.
[31] P. Bakshi, D. A. Broido, and K. Kempa, *Phys. Rev.* **B 42** (1990) 7416.
[32] P. L. McEuen, N. S. Wingreen, E. B. Foxman, J. Kinaret, U. Meirav, M. A. Kastner, and Y. Meir, *Physica* **B 189** (1993) 70.
[33] U. Merkt, *Physica* **B 189** (1993) 165.
[34] P. A. Maksym and T. Chakraborty, *Phys. Rev. Lett.* **65** (1990) 108.
[35] P. A. Maksym and T. Chakraborty, *Phys. Rev.* **B 45** (1992) 1947.
[36] P. A. Maksym, *Physica* **B 184** (1993) 385.
[37] Partoens, A. Matulis, and F. M. Peeters, *Phys. Rev.* **B 59** (1999) 1617.
[38] H. Imamura, P. A. Maksym, and H. Aoki, *Phys. Rev.* **B 53** (1996) 12613.
[39] Pfannkuche and R. R. Gerhardts, *Phys. Rev.* **B 44** (1991) 13132.
[40] V. Gudmundsson and R. R. Gerhardts, *Phys. Rev.* **B 43** (1991) 12098.
[41] Z. L. Ye and E. Zaremba, *Phys. Rev.* **B 50** (1994) 17217.
[42] M. Peeters and V. A. Schweigert, *Phys. Rev.* **B 53** (1996) 1468.
[43] Brataas, U. Hanke, and K. A. Chao, *Semi. Sci. Tech.* **12** (1997) 825.
[44] Matulis, J. O. Fjarestad, and K. A. Chao, *Physica Scripta* **T69** (1997) 85; T69 (1997) 138.
[45] Anisimovas and A. Matulis, *J. Phys.: Condens. Matter* **10** (1998) 601.
[46] N. Akman and M. Tomak, *Physica* **B 262** (1999) 317.
[47] K. Jauregui, W. Hausler, and B. Kramer, *Europhys. Lett.* **24** (1993) 581.
[48] X. -G. Li, W. -Y. Ruan, C. -G. Bao, and Y. -Y. Liu, *Few-Body Systems* **22** (1997) 91.
[49] Szafran, F. M. Peeters, S. Bednarek, T. Chwiej and J. Adamowski, *Phys. Rev.* **B 70** (2004) 035401.
[50] J. H. Jefferson, W. Hausler and D. Malvern, *Mol. Phys. Rep.* **17** (1997) 81.
[51] C. Thompson and A. Alavi, *Phys. Rev.* **B 69** (2004) 201302R.
[52] Reusch and H. Grabert, *Phys. Rev.* **B 68** (2003) 45309.

[53] P. A. Sundqvist, S. Volkov, Y. E. Lozovik and M. Willander, *Phys. Rev.* **B 66** (2002) 75335.
[54] R. Egger, W. Hausler, C. H. Mak and H. Grabert, *Phys. Rev. Lett.* **82** (1999) 3320.
[55] Tanatar and D. M. Ceperley, *Phys. Rev.* **B 39** (1989) 5005.
[56] S. T. Chui and B. Tanatar, *Phys. Rev. Lett.* **74** (1995) 458.
[57] S. A. Mikhailov, *Phys. Rev.* **B 65** (2002) 115312.
[58] N. Akman and M. Tomak, *Physica* **E 4** (1999) 277.
[59] Brataas, U. Hanke, and K. A. Chao, *Phys. Rev.* **B 54** (1996) 10736.

In: Frontiers in Quantum Dots Research
Editor: Marco L. Voronin, pp. 85-103
ISBN: 1-60021-057-0

Chapter 4

RESEARCH PROGRESS IN II-VI QUANTUM DOTS

***C.X. Shan*[1] *and X.W. Fan*[2]**

[1] School of Physical Engineering and Laboratory of Material Physics, Zhengzhou University, Zhengzhou, 450052, China.

[2] Key Laboratory of Excited State Processes, Changchun Institute of Optics, Fine Mechanics and Physics, Chinese Academy of Sciences, Changchun, 130033, P.R. China

Abstract

Quantum dot (QD) structures have attracted considerable attention in the past decades because they not only address an important issue in basic physics, but also find potential and practical applications in many fields. II-VI materials have long been studied for their larger band gap, higher exciton binding energy and larger polarity than their III-V counterparts. If combining the advantages of QD structures and those of II-VI materials, a charming prospect can be expected. Under this circumstance, many groups began to study the II-VI QDs. Some noteworthy progresses have been obtained after several years' research. In this paper, we will summarize these progresses, and the summarization will be divided into five parts: 1. The preparation of II-VI QDs. In this part, we will demonstrate the two epitaxial growth modes of QDs, Stranski-Krastanow (S-K) mode and Volmer-Weber (V-W) mode. 2. The formation mechanism of S-K mode QDs. We depict the formation mechanism of QDs grown in S-K mode, in which the energy reduction induced by atom diffusion will be considered. 3. The optical properties of II-VI QDs. 4. Progress in the practical applications of II-VI QDs. The light emitting devices and lasing diodes based on CdSe, CdS QDs will be summarized in this part. 5. The prospect of II-VI QDs. In this part, the unresolved problems in II-VI QDs are addressed, and their possible solutions are also dealt with.

1. Introduction

Quantum dot (QD) is very small system with its three-dimensional (3 D) sizes comparable or smaller than the de Broglie wavelength of electrons. Having 3 D confinement to carriers in them, QD structures have attract considerable attention in the past decades because they are of great significance in both basic physics and practical applications [1, 2]. Semiconducting

QDs attract particular attention for their possible integration with existing micro- and optoelectronic technologies. In 1982, Arakawa and Sakaki predicted that lasers based on semiconductor QD structures had many advantages over those on quantum well (QW) structures in many aspects, such as lower threshold current density, higher gain and efficiency, better modulation properties and temperature stability, etc [3]. Besides the application in lasers, semiconductor QDs have found potential as well as practical applications in storage devices [4], detectors [5, 6], and quantum computation [7].

As one important portion of semiconductor, II-VI materials, such as ZnS, ZnSe, CdS, CdSe have long been studied for their larger band gap, higher exciton binding energy and larger polarity than their III-V counterparts. Significant progresses have been achieved in II-VI light-emitting-device (LED) and laser diode (LD) [8, 9] in the early 1990's. In recent years, the application in white light emission of ZnSe-base materials has been reported [10, 11]. Especially, the booming of ZnO in the near past agitates the study on II-VI materials greatly. Attracted by the strong points of II-VI materials and the superiorities of QD structures, many groups began to study the II-VI QDs. The earliest II-VI QD structures are CdS, CdSe and their alloy in nanometer scale dispersed into silica, polymer and other transparent materials [12]. In 1930's, there were reports on such material systems, while until 1980's, researchers studied this kind of glass in terms of quantum size effect [13]. Later the study on II-VI QDs was concentrated on nanocrystal or colloid [14, 15]. These two kinds of QDs are still charming at present. In 1996, Xin *et. al.* prepared CdSe QDs by molecular beam epitaxy (MBE) [16], this is the first report on "modern" II-VI QDs. Significant progresses have been obtained after about ten years' research. In this section, we will summarize the recent research progress in this field.

2. Fabrication of II-VI QDs

There are two kinds of methods employed to the fabrication of QD structures, patterning and epitaxial route. The patterning route is to prepare a patterned substrate in mechanical or chemical way, and the QDs were then selectively grown on the substrate. The advantage of this method is that it can produce dots with controlled density, size and alignment. Figure 1 shows the ZnCdS QDs prepared in this method [17], indicating the uniform in size and good alignment of the dots. Kim *et. al.* also prepared ZnO nanodots in a Ga^+ patterned SiO_2/Si substrate by metalorganic chemical vapor deposition (MOCVD) technique [18]. The shortcoming of this method is that the size of the dots is restricted by the processing resolution, and the chemical or mechanical processing to the substrate will inevitably impair the crystal quality of the QDs. Consequently, the characteristics of the devices based on QDs prepared in this route are far from expectation. The epitaxial route avoids the artificial processing to the material system, and the QDs form by itself. Since this method doesn't introduce intentional damnifications to the dots, it is widely used to prepare high quality QD structures. This section will mainly concentrate on II-VI QDs prepared in this method.

The epitaxial growth by MOCVD and MBE is a non-equilibrium process, and QDs fabricated in this method can be categorized into two different modes, Stranski-Krastanow (S-K) and Volmer-Weber (V-W) mode [19, 20]. The schematic illustration of the two growth modes was demonstrated in Fig. 2. S-K mode is usually applied in systems with large lattice mismatch. In the initial stage of S-K mode, the material grows layer-by-layer. Due to the large mismatch between the epilayer and substrate, there will be strain in the epilayer, and the strain will increase with the thickness of the two-dimensional (2 D) layer. When the thickness of the 2 D layer reaches a certain value that is entitled critical thickness, the 3 D dots form on the 2 D layer to release the strain, and the 2 D layer lying beneath the dot is called wetting layer. Contrary to S-K mode, in V-W mode, the large lattice mismatch is not necessary, 3 D dots form directed on the substrate without the 2 D layer-by-layer growth stage, and the dots grow in size with increasing the precursor supplement.

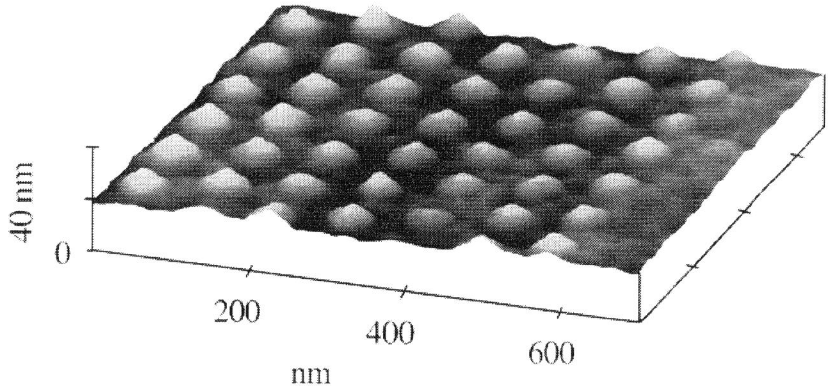

Figure 1. AFM image of ZnCdS–ZnMgCdS heterostructure QDs selectively grown on GaAs [17].

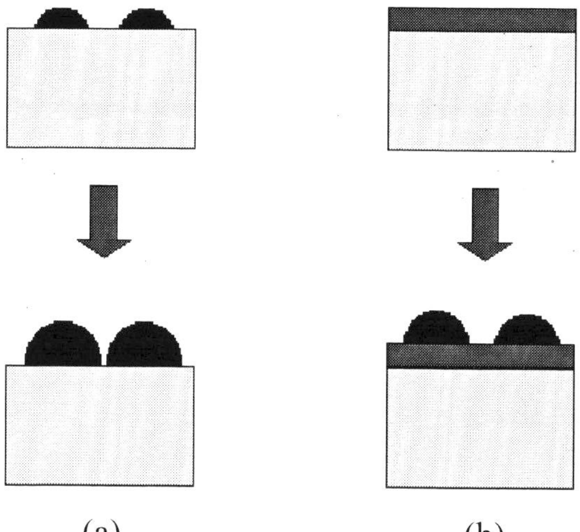

Figure 2. The two modes corresponding to the formation of QDs, (a) V-W mode; (b) S-K mode [18, 19].

2.1. II-VI QDs Grown in S-K Mode

Xin *et. al.* prepared CdSe/ZnSe QD structures by MBE technique using the InAs/GaAs QDs for reference, for these two systems have similar lattice mismatch and band gap discontinuity [16]. They first deposited a 2 μm ZnSe buffer layer on GaAs substrate, then a CdSe layer with several monolayers (ML) thick was deposited on the ZnSe layer. After the depositing of the CdSe layer, the reflection high-energy electron diffraction (RHEED) pattern changes from streaky to spotty, indicating the growth mode changes from 2 D to 3 D. The sample was taken out of the growth chamber when the growth is completed, and its morphology was investigated by atomic force microscopy (AFM). Figure 3 shows the AFM image of the CdSe/ZnSe system. As shown in the figure, there appear small dot-like structures on the ZnSe surface. Since the transition from 2 D to 3 D growth mode has been observed in the formation process of the dots, they considered that the dots were formed in S-K mode. The reports on II-VI QDs mushroomed after that, and QDs have been realized in some other systems with large lattice mismatch, such as ZnSe/ZnS [21, 22], CdTe/ZnTe [23, 24], and CdSe/MgS [25].

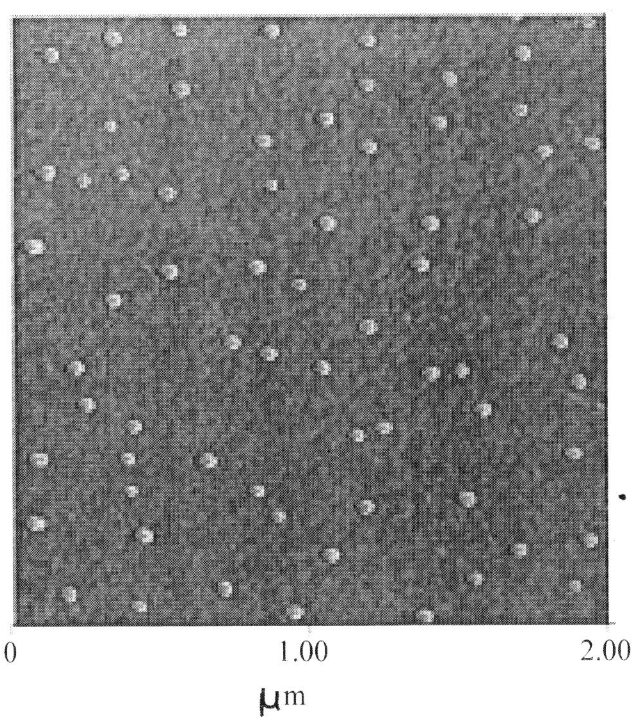

Figure 3. AFM image of CdSe QDs grown on ZnSe [16].

As for the formation dynamics of the dots, Kim *et. al.* observed the images of the 1.5 ML CdSe deposited on ZnSe by scanning transmission electron microscopy (STEM) [26]. They found the coexistence of the 2 D platelets and the 3 D islands, which clearly indicates that the platelets act as the precursors for the islands. Their results reveal the transition process from the 2 D epilayer to the 3 D islands.

We have prepared $Zn_{0.56}Cd_{0.44}Se$ alloy QD structures in S-K mode by MOCVD, and the formation process of the dots was studied by PL measurements [27]. Prior to the growth, the critical thickness required for the formation of QDs was calculated in terms of the strain relaxation theory, then the growth was guided by the calculated results. When the $Zn_{0.56}Cd_{0.44}Se$ is 3.0 nm (almost equal to the calculated critical thickness of 3.1 nm), both atomic force microscopy and PL measurements indicate the formation of QD structure. Based on the fact that the dots have different PL spectra at different formation stage, we studied the formation process of QDs by performing PL measurement on a series of ZnCdSe/ZnSe quantum structures with different ZnCdSe thickness, as shown in Fig.4. When the thickness of ZnCdSe is 1.8 and 2.1 nm, the room temperature PL spectrum shows one Gaussian lineshape, as indicated by Curve A and B. While when the thickness is 2.5 nm, the PL spectrum shows an asymmetry shape, and it can be simulated into two Gaussian lines, marked by Peak I and II. The peak at higher energy is believed to come from the 2 D wetting layer, and the one at lower energy come from the quasi-3 D dots. The 3 D dots can be formed below the critical thickness, but that the size of the dots is small, and the density is relative low [28, 29]. This may be the reason for the small intensity ratio of Peak I to Peak II. When the thickness of the ZnCdSe layer is 3.0 nm, as shown by Curve D in Fig. 4, much more 3 D dots appear because the thickness is very close to the calculated critical thickness, significant mass transfer from the 2 D wetting layer to the QDs occurs. Consequently, the size of the 3 D dots increases at the cost of the 2 D wetting layer. Therefore, Peak I from the 3 D dot redshifts, while Peak II from the 2 D layer blueshifts due to the attenuation of the wetting layer. When the ZnCdSe thickness is increased further, for instance, to 3.6 nm, Peak I shows slight redshift, while Peak II redshifts drastically compare to Curve D. By comparing the emission peak of the QDs for Curve D with that for Curve E, one can deduce that the increase in dot size is very small when the additional 0.5 nm of ZnCdSe was delivered, and most of the excrescent materials are used to increase the thickness of the wetting layer.

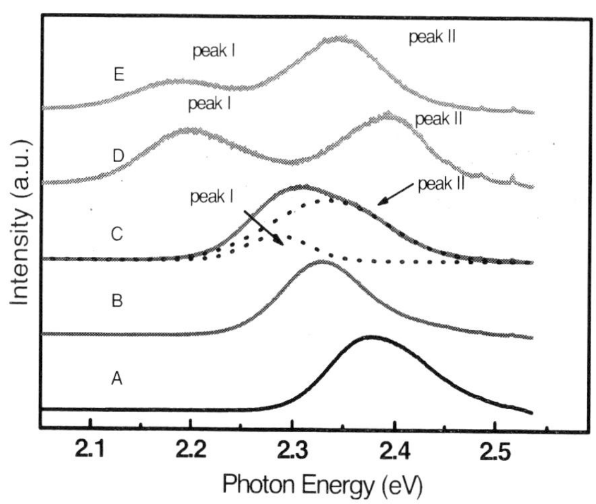

Figure 4. PL spectra of ZnCdSe/ZnSe quantum structure with different ZnCdSe thicknesses at room temperature. The thicknesses are (A) 1.8 nm, (B) 2.1 nm, (C) 2.5 nm, (D) 3.0 nm, (E) 3.6 nm [27].

The formation of highly correlated vertical QD stacks is considered a significant step to the practical application of QD structures. However, due to the suppleness of II-VI materials, there is a competition between island formation and energy relaxation via misfit dislocations in the formation process of S-K mode QDs, which hinders the formation of II-VI QD stacks. Kurtz *et. al.* prepared vertically stacked CdSe based QD structure by a modified MBE technique [30]. They employed CdS compound as the Cd source instead of the conventional Cd elemental source, and in this way, the highly correlated vertical CdSe QD stacks had been obtained. The TEM image of the stacks was shown in Fig. 5, which is very similar to the previous reported InAs/GaAs QD stacks [31].

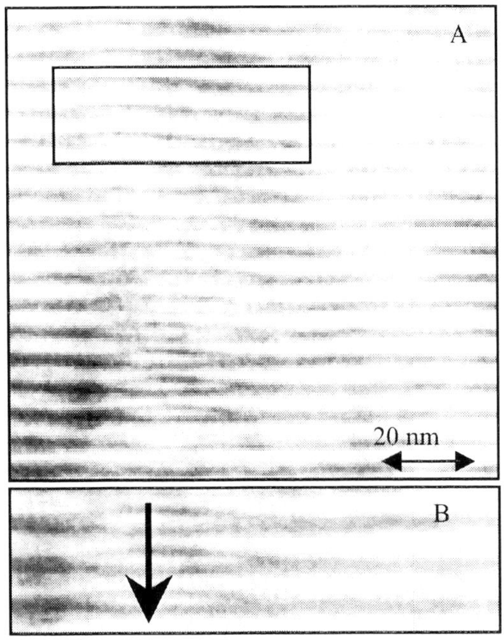

Figure 5. (A) Cross-section TEM image of the CdSe/ZnSe QD stacks. (B) The magnification image of the area in the pane in (A), and the arrow clearly indicates the stacking of the dots [30].

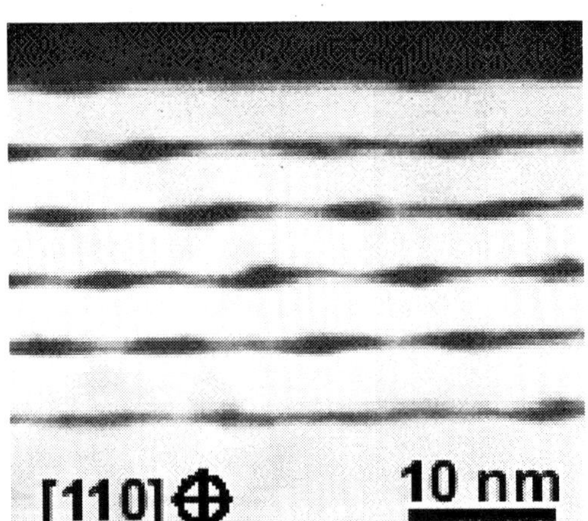

Figure 6. TEM image of CdSe quantum dots arrays grown on BeTe with CdTe as strain modulator [34].

Ivanov *et. al.* developed a novel method to fabricate well aligned II-VI QDs, which is named stressor-controlled epitaxy [32-36]. Different from the conventionally employed S-K mode, in this method, a super-strained fractional ML material that has a larger lattice mismatch with the substrate was introduced. As a result, a stronger stress field was generated, and dots form under the modulation of this stress field. The fractional material is called "stressor". The stressor material is deposited on the substrate before the deposition of the basic QD material. Figure 6 shows the TEM image of the CdSe/BeTe QD structure prepared by Ivanov *et. al.* [34]. CdTe was chose as the stressor since it has a lattice mismatch of as large as 15% with the BeTe layer. Thus, a regular stress field across the BeTe surface was introduced by alternating regions with CdTe and BeTe bonds. This results in a vertical chess-ordering in the CdSe/BeTe QD multilayers, as observed in the cross-sectional TEM image in Fig. 6.

2.2. II-VI QDs Grown in V-W Mode

The large lattice mismatch required in S-K mode is not necessary in V-W mode QDs, and the growth parameters are not so strict as those in S-K mode. Therefore, for its simplification in growth conditions, the QDs grown in this mode attract some attention.

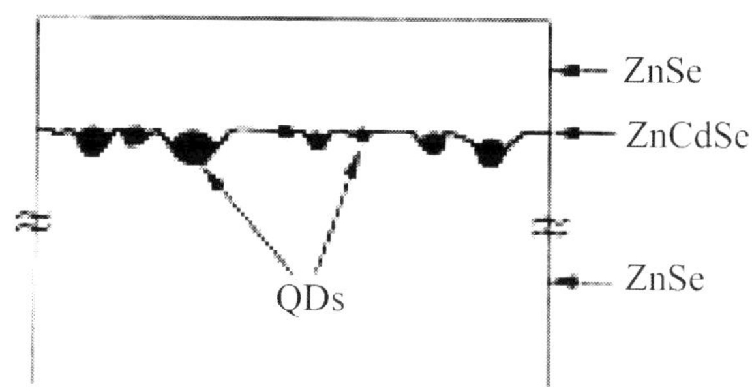

Figure 7. Schematic diagram of ZnCdSe quantum dots on a rough ZnSe surface [40].

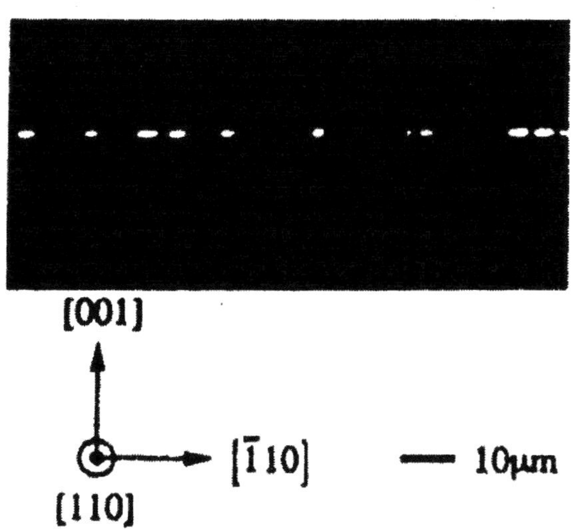

Figure 8. Monochromatic CL image taken at 5 K with a wavelength of 450 nm [38].

In 1997, Zhang *et. al.* prepared ZnCdSe QDs by MBE in V-W mode [37-40]. They first grew a ZnSe layer on cleaved ZnSe (110) surface. When the atom ratio of Zn: Se is not 1: 1, the ZnSe layer will not be grown layer-by-layer, and fluctuations in thickness will appear. By depositing a several monolayer ZnCdSe layer on the fluctuated ZnSe layer, whose band gap is smaller than ZnSe, island structures will form. By capping the islands with ZnSe, ZnCdSe

QDs was then prepared. The schematic diagram of the QD structure prepared in this way was shown in Fig. 7. Figure 8 is the 5 K monochromatic cathodoluminescence (CL) image of the ZnCdSe/ZnSe structure taken at a wavelength of 450 nm. If the prepared structure is quantum well, a uniform stripe will appear in the CL image. The continuous dot-like image clearly indicate that the formation of 3 D QD structures.

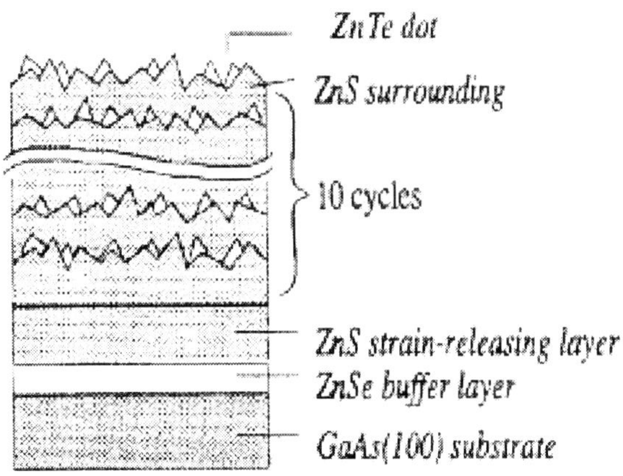

Figure 9. Schematic illustration of ZnTe/ZnS QD structure grown in V-W mode [42].

Chen *et. al.* in National Taiwan University have devoted to the preparation of QD structures in V-W mode by MOCVD technique [41-43]. They grew a material with large band gap, ZnS for example, on substrate with a fast growth rate. Then the surface of the ZnS layer will have many fluctuations for the fast growth rate. Then a layer with smaller band gap, ZnSe in this case, was deposited on the ZnS layer. Repeat the above-mentioned process several times, they got the ZnSe QD structures. ZnTe dots were also prepared in this way. The schematic illustration of the structure of ZnTe QW was demonstrated in Fig. 9.

3. The Formation Mechanism of S-K Mode QDs

One of the most important issues in QDs is that the realization of the controlled growth of the dots. Therefore, exploring the formation mechanism of QDs, so as to understand the key parameters in forming dot structure, is of great significance. Seifert *et. al.* had studied the formation mechanism of nanostructures [44], and they considered that the mechanism for the formation of S-K QDs was as that shown in Fig. 10. In this figure, the horizontal abscissa is time and the vertical one is the total energy of the investigated system. The accumulated strain induced by mismatch increases almost linearly with the thickness of the epilayer in Region A. In this region, the epilayer grows in stable layer-by-layer 2 D mode when its thickness is smaller than t_{cw}, while grows in a metastable mode when larger. At point X, the growth mode of the epilayer changes from metastable 2 D to 3 D in a very short time span, and the energy released by forming 3 D islands in Region B. The energy E_A is the energy barrier for the growth mode transition. The formed islands will interact each other to decrease the energy of the system further, the process is called the ripening process, as indicated by Region C.

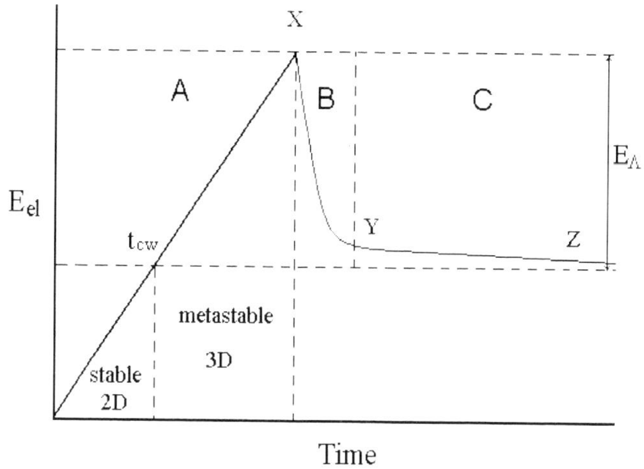

Figure 10. Schematic of total energy versus time for the 2 D- 3 D morphology transition. A: 2 D layer-by-layer growth; B: 2 D- 3 D transition; C: ripening process. E_A is the barrier for formation 3 D islands, X is the point where a pure strain-induced transition becomes possible. Between Y and Z a slow ripening process continues [44].

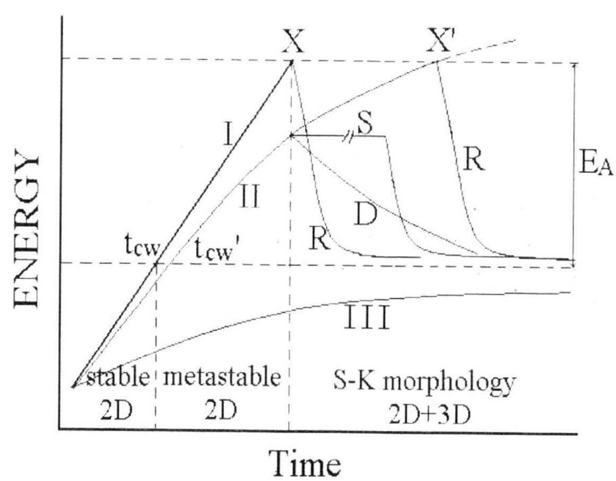

Figure 11. The formation mechanism of S-K mode QDs modified by Yang et. al. [46].

The mechanism depicted above is applicable in ideal system, while in "real" system, it is something different. Many researchers found the formation of QDs below the critical thickness [28, 29], which is contrary to the mechanism. Yang *et. al.* studied the CdSe/ZnSe system with different CdSe thickness [45]. They thought that the CdSe layer would have fluctuations in thickness inevitably, the adsorbed atoms or radicals on the substrate surface will move under the action of heat energy, and the movement will impact the formation of the dots greatly. Therefore, they made some modifications to the above-mentioned mechanism, as shown in Fig. 11 [46].

In the real growth process, some of the total energy of the system will be released by the movement of atoms on the surface. Thus in Region A, the total energy marked by curve II is slightly slower then the theoretical value of curve I. Therefore, to get across the barrier of

E_{EA}, a thicker 2 D growth layer is needed. Then the dots form through Process R. Another case is that the 2 D layer is not thick enough, and the total energy of the system lies between t_{cw}' and X'. Then the strain will be released via a competing mode (process S and D). In Process D, the strain was released by forming dislocations. While in Process S, since the thickness is below the critical thickness that is required for the formation of QDs, 3 D dots will not form immediately. However, the movement of the atoms or radicals on the surface will lead to some areas reach or excess the critical thickness occasionally, then 3 D dots formed there. The dots form under the assistance of atom movement requires a much longer time span than those form directly. In some cases, most of the strain was released by the atom movement, and the 3 D islands will not form in the entire growth process, as shown in curve III.

4. Optical Properties OF II-VI QDs

To realize the practical application of QDs, good optical properties are usually required. Therefore, the studies on the optical properties have been charming since the discovery of QDs. Usually, an ensemble of dots show inhomogeneous broadening in emission due to the wide distribution of size, shape and surrounding environment of the dots. While single dot spectroscopy can reveal the optical spectrum of an individual dot free of the interferences of the distributions. Kim *et. al.* studied the temperature dependent PL spectrum of the CdSe QDs by using micro- and nano-PL spectroscopy [47]. They studied the emission from both the individual dots and the QD ensemble by changing the spatial resolution. They found that at low temperature, the spectrum was composed by sharp lines and broad emissions, and the sharp lines superimposed on the broad emission.

Figure 12a shows the nano-PL spectra of the sample as a function of temperature from 7 to 60 K. There appears a sharp peak in each spectrum, and it can be assigned the emission from an individual dot. The width of the dot broadens from 200 μeV at 7 K to 900 μeV at 60 K, and the intensity decreases gradually. By monitoring the quenching of the single-dot emission with temperature, an activation energy of 4.0±0.7 meV can be deduced from the spectra. For the broad emission, the activation energy is about 40 meV. Therefore they argued that two different states may be responsible for the emission of the QD sample. One is the sharp lines originating from the individual dots, the other is the broad emission.

One of the most important applications of QDs is that they can be used as the single photon source. Sebald *et. al.* have reported the generation of triggered single photons obtained from S-K mode CdSe/Zn(S, Se) QD system at temperature up to 200 K [48]. Figure 13 is the μ-PL spectra of a separated exciton emission line. At 4.5 K the emission peak exhibits a Lorentzian profile with a linewidth of 0.5 meV. With increasing temperature, sidebands appear and eventually become dominant. For temperatures above 80 K the phonon sideband of an emission line located at an energy of 2.392 eV originating from another QD starts to overlap with the exciton emission line at 2.369 eV. The spectral linewidth increases with increasing temperature eventually. Their results demonstrate the potential of self-assembled CdSe/Zn(S, Se) QDs for use in non-classical light generation at temperatures up to 200 K. Later, Ulrich *et. al.* reported the triggered polarization –correlated photon pairs from a single CdSe QDs [49].

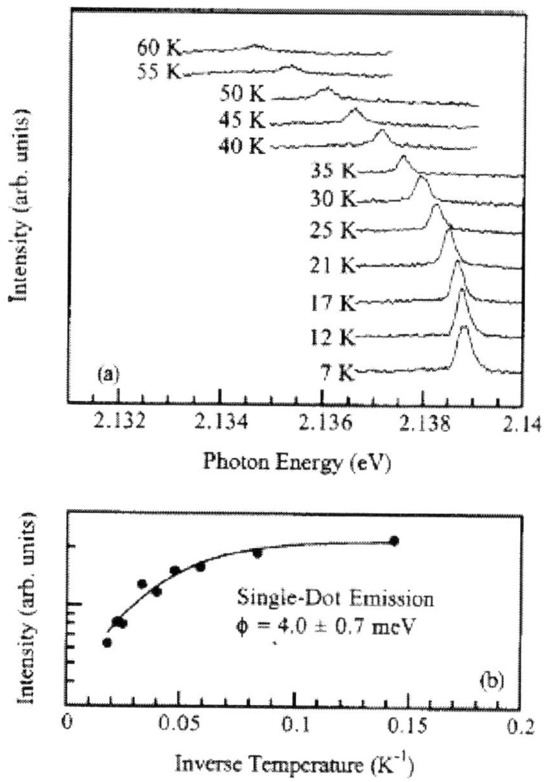

Figure 12. (a) PL measurements of a single dot as a function of temperature; the data were obtained through a 200 nm fixed aperture. (b) The spectrally integrated intensity of the quantum dot PL is plotted vs inverse temperature; this data yields an activation energy for the quantum dot of 4.0±0.7 meV [47].

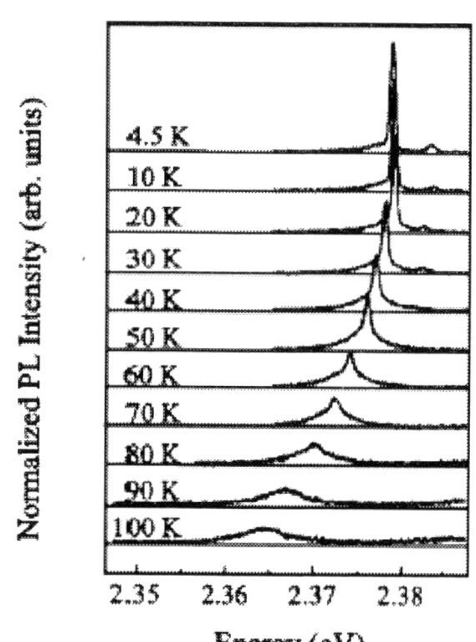

Figure 13. The μ-PL spectra of the CdSe QD sample excited at 435 nm (P=3.3 kW/cm^2) at different temperatures. The spectra are normalized to the integrated intensity at 4.5 K and shifted vertically for better representation [48].

As for the newcomer ZnO QDs, Fonoberov and coworkers have studied the origin of the ultraviolet PL from the ZnO dots in theory. They found that there were two possible sources for the ultraviolet emission, excitons confined in the QDs and excitons bound to ionized impurities located at the QDs surface. The origin of the ultraviolet PL in ZnO dot can be attributed to the either one depending on the fabrication method and surface passivation technique. The two origins can be distinguished by measuring the exciton radiative lifetime [50].

5. Progress in the Practical Applications of II-VI QDs

The above-mentioned preparation methods, formation mechanism, and optical properties of II-VI QDs lay a good foundation for the potential application of them. Compare to the significant progress in III-VI QDs based devices [2, 51], the steps in II-VI QDs devices are relative small. However, attracted by the unique properties of II-VI compounds, many researchers devoted themselves to the research on LED and LD based on II-VI QDs, and some noteworthy results were obtained.

Some researchers in Department of Electronics and Mechanic Engineering, Chiba University studied CdS QDs since 1997 [52- 56]. The first CdS LED was reported in 1999 by them. They prepared CdS dots sandwiched in the *p-n* junction of ZnSe layer grown on GaAs(100) substrate. Bright blue or green luminescence induced by current injection was observed at 77 K as well as room temperature. The schematic of the CdS QDs LED structure was shown in Fig. 14. The carrier concentration in *n*-ZnSe in this structure is 5×10^{17} cm^{-3}, and the dopant concentration in *p*-ZnSe is N_a- N_d~ 5×10^{17} cm^{-3}. Blue and green emissions were obtained from this structure, as shown in Fig. 15.

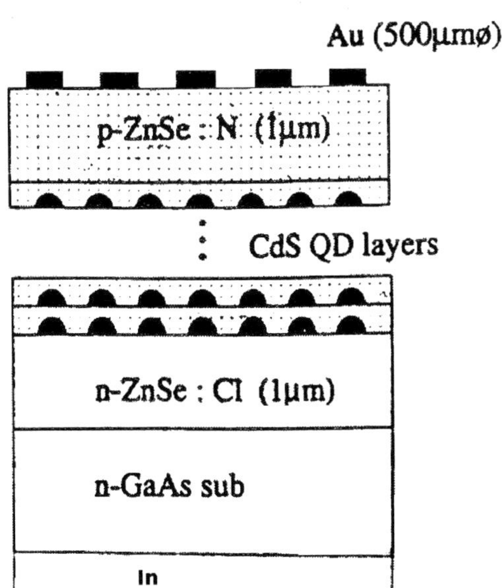

Figure 14. Schematic of the device structure of CdS QDs LED [53].

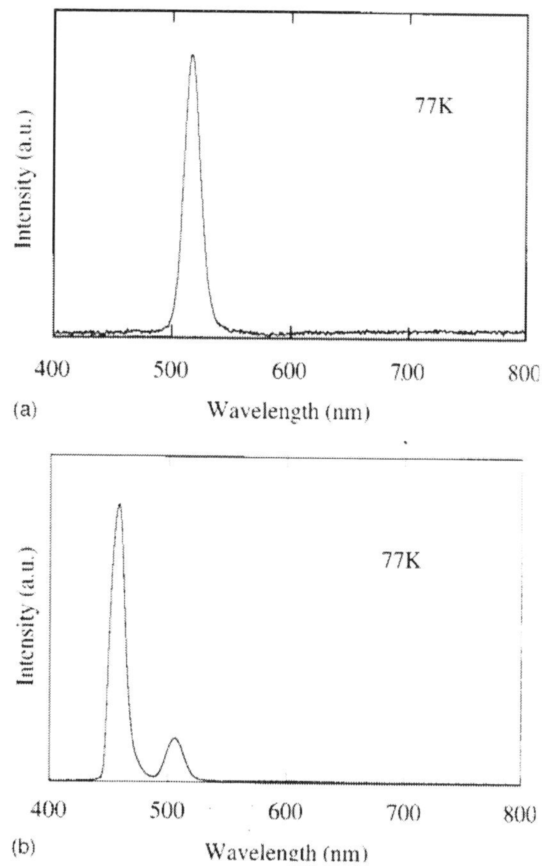

Figure 15. Green (a) and Blue (b) EL spectrum of CdS QDs LED [55].

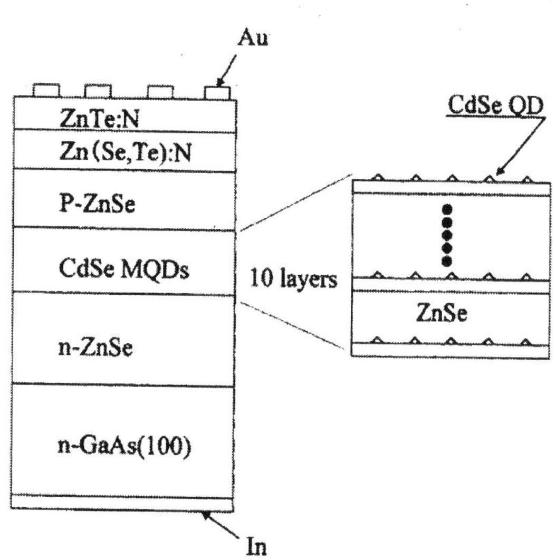

Figure 16. Structure of ZnSe p–n diode with CdSe multi QD layers [58].

Passow *et. al.* reported the electroluminescence (EL) of CdSe QD structures in 2000 for the first time [57]. In 2002, researchers grew CdSe multiple QD layers between *n* and *p*-ZnSe layer, and produced the CdSe QDs LED [58]. The illustration of the LED structure is shown in Fig.16. Green current injection induced emission was observed to the naked eye in a dark room. While white emission was observed in ZnSe *p-n* diodes without CdSe quantum dots. Therefore, they judged that the observed green electroluminescence originated from the CdSe quantum dots. The lifetime of the obtained LED is more than 10 h at 77 K.

The most prominent merits of QDs lie in the priority in LD structures based on it. Therefore, the study on QD-base LD is much more attractive. We have mentioned in section 2.1 that Ivanov *et. al.* prepared CdSe QD structure in a strain-modulated method [30- 34, 59]. The photo-pumped lasing was observed in this structure with the threshold excitation intensity of about 4 kW/cm^2. The threshold excitation intensity of CdSe QD and ZnCdSe QW structures versus temperature is shown in Fig. 17, the inset shows the PL spectra of the QD structure under and above the threshold excitation intensity. As shown in the figure, much narrower but stronger emission was observed when the excitation intensity is 3.9 kW/cm^2, and the temperature stability of the QD structure is much better than the QW one, just as that predicted by Arakawa and Sakaki [3]. They also prepared LD structures employing CdSe QDs as the active layer. The room temperature EL spectra of this structure were shown in Fig. 18. The spectrum shew a broad line-shape when the injection current is 150 mA, while when it is 500 mA, a very sharp and strong peak appeared, indicating lasing action was realized. However, the laser diode is not stable, as shown in the inset, the peak position shifts with time.

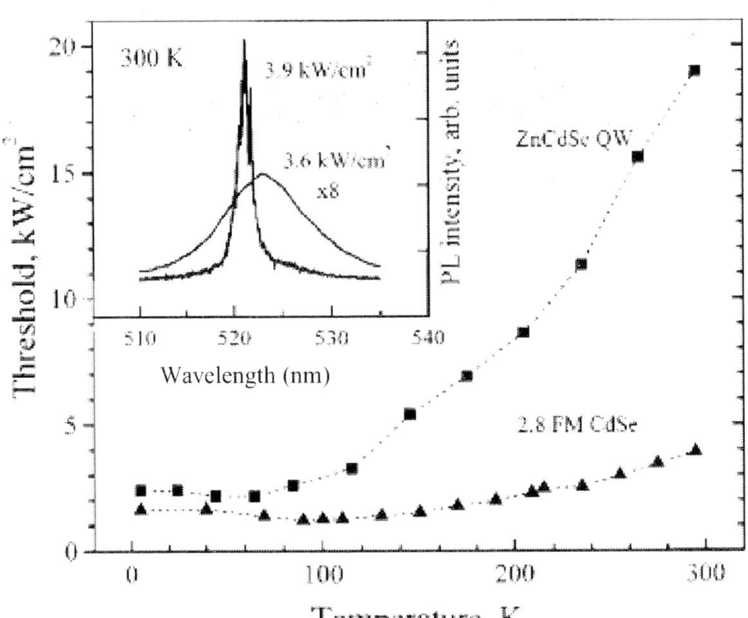

Figure 17. Temperature dependence of threshold power density for QDs (▲) and QWs (■) laser structures. Room temperature FM laser generation spectra are presented in the inset [59].

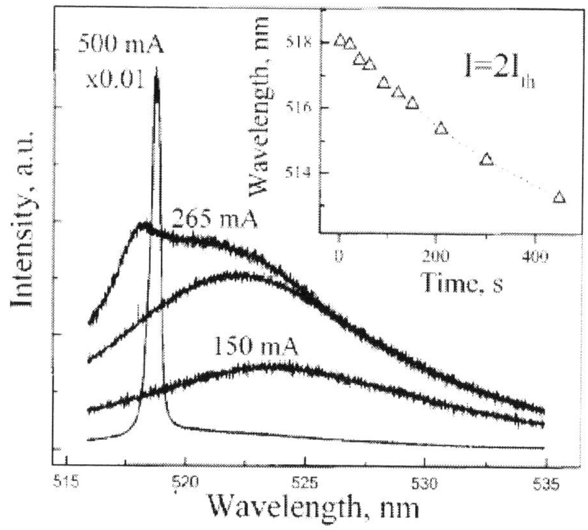

Figure 18. Room temperature electroluminescence spectra of CdSe QDs based laser diode. The time dependence of lasing wavelength is shown in the inset [59].

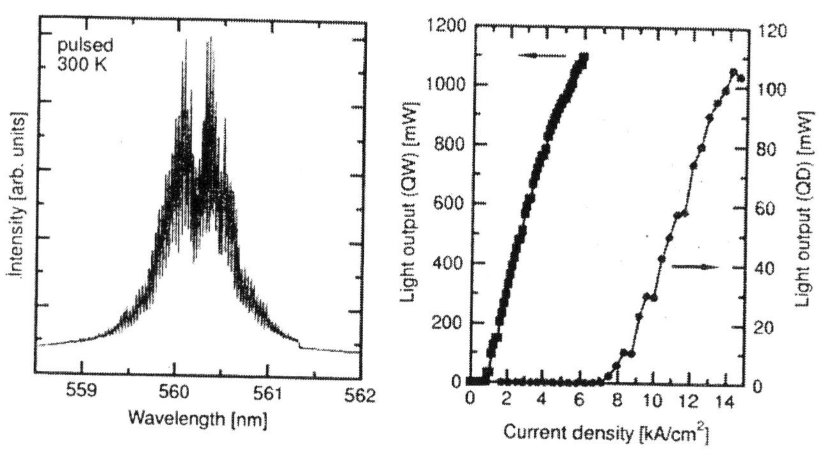

Figure 19. Lasing characteristics of the quantum dot laser. Left: Laser spectrum recorded in pulsed mode. Lasing around 560 nm is realized. Right: L-J characteristics for quantum dot and quantum well laser [60].

Klude *et. al.* also realized the current injected lasing action of CdSe QDs in 2002 [60]. Figure 19 is the lasing characteristics of the diode. The emission wavelength is 560 nm with the threshold current density of 7.5 kA/cm^2. The right hand part of the figure shows the corresponding light output versus current density characteristics. The threshold current density of the QDs LD is almost one order higher than that of the ZnCdSe QW one. Despite the high threshold, a maximum output power of more than 100 mW per pulse can be obtained in the QDs LD structure. The devices can operate up to more than 14 kA/cm^2, which has never been reported for ZnSe-based devices before. With a higher volume of active material (i.e. more CdSe sheets) and a narrower size distribution of the dots, a significant reduction of the threshold current density can be expected. The lifetime of the devices is around one minute in pulsed mode which is significantly shorter than the QW based LD. The high threshold current densities may be responsible for the short lifetimes. By optimizing the structural parameters of the QDs LD, the operation of the structure may be greatly improved.

6. Prospect of II-VI QDs

Although some progresses have been achieved in II-VI QD structures as indicated above, many issues are still under debate in basic physical problems of them, and the progress in practical applications still has a huge gap with their III-V counterparts. The unresolved problems are listed below:

1. Size distribution of QDs.

One of the most prominent characters in QDs is their atom-like discrete energy level. However, the luminescence spectrum of QDs usually presents itself in a broad line-shape, and the separated peaks can not be recognized at room temperature. This is mainly due to the large size distribution of the dots. The broad size distribution tampers the unique characters of the QD structure, and it is an impediment to the QDs laser diodes [61]. To realize practical application in diodes, the size fluctuation less than 5- 15% is required in theory [62]. However, the fluctuation of QDs obtained at present is usually at 15- 50% [62], which is much larger than the theoretical value. Therefore, the realization of QDs with uniform size has long been one of the most important issues in II-VI QDs.

2. Stability of II-VI QDs

Another character of II-VI QDs is the ripening, namely the dots evolve in density and size with time. Generally speaking, the density of the dots decreases and the size increases with time elapsing. The instability of the dots has been a topic in heat debate since the first II-VI QDs were prepared in 1996 [16]. Some researchers think that the ripening process is resulted from the oxidation of selenium [63], while others refer it as the intrinsic properties of QDs [64]. Although numerous studies have been carried out on the ripening process of II-VI QDs, this issue has not been well solved yet.

3. *P*-doping of II-VI QDs

II–VI QDs prove to be important for the future applications as well as the basic physics. *P*-type doping is a key issue for their practical device applications but is the long-lasting issue to be solved. Two novel methods were put forward recently to realize the *p*-doping of II-VI materials, one is codoping method, and the other is the doping of nano-clusters which are heavily *p*-doped. Hole concentrations in the range of 1×10^{17} to 1×10^{18} cm^{-3} were obtained without post-growth annealing by applying the latter method [59].

In summary, although some progresses have been achieved in the preparation, formation mechanism, optical properties, and potential applications in II-VI QDs, there is still a lot of work to do in this field. In recent years, the mushrooming of ZnO QDs agitates the research of II-VI QDs further [65, 66]. In addition, the successful fabrication of QDs LED with much higher efficiency using CdSe nanocrystals sandwiched by two organic films recently may promise an inspiring prospect of the II-VI QDs [67, 68].

References

[1] Nőtzel R.; Emmyo J.; Tamamura T. *Nature*, 1994, 369, 131- 133.

[2] Liu H. Y.; Xu B.; Wei Y. Q.; Ding D.; Qian J. J.; Han Q.; Liang J. B.; Wang Z. G. *Appl. Phys. Lett.* 2001, 79, 2868- 2870.

[3] Arakawa Y.; Sakaki H. *Appl. Phys. Lett.* 1982, 40, 939- 941.

[4] Bodefeld M. C.; Warburton R. J.; Karrai K.; Kotthaus J. P.; Ribeiro G. M.; Petroff P. M. *Appl. Phys. Lett.* 1999, 74, 1839- 1841.

[5] Soref R. *MRS Bulletin* 1998, 23, 20- 24.

[6] Schittenhelm P.; Engel C.; Findeis F.; Abstreiter G.; Darhuber A. A.; Bauer G.; Kosogov A. O.; Werner P. J. *Vac. Sci. Technol. B* 1998, 16, 1575- 1581.

[7] Bonadeo N. H.; Erland J.; Gammon D.; Park D.; Katzer D. S.; Steel D. G. *Science* 1998, 282, 1473- 1476.

[8] Haase M. A.; Qiu J.; Depuydt J. M.; Cheng H. *Appl. Phys. Lett.* 1991, 59, 1272- 1274.

[9] Kato E.; Noguchi H.; Nagai M.; Okuyama H.; Kijima S.; Ishibashi A. *Electron. Lett.* 1998, 34, 282- 284.

[10] Luo Y.; Guo S.P.; Maksimov O.; Tamargo M.C.; Asnin V.; Pollak F.H.; Chen Y.C. *Appl. Phys. Lett.* 2000, 77, 4259- 4261.

[11] Luo Y.; Elmoumni A.; Guo S. P.; Tamargo M. C.; Kelly S.; Ghaemi H., Asnin V.; Tomkiewicz M.; Pollak F. H.; Chen Y. C. J. *Vac. Sci. Technol. B* 2000, 18, 1522- 1525.

[12] Inokuma T.; Arai T.; Ishikawa M. *Phys. Rev. B* 1990, 42, 11093- 11098.

[13] Rosetti R.; Hull R.; Ellison J.L.; Brus L.E.; *J. Chem. Phys.* 1984, 80, 4464- 4469.

[14] Peng X. G.; Manna L.; Yang W. D.; Wickham J.; Scher E.; Kadavanich A.; Alivisatos A. P. *Nature* 2000, 404, 59- 61.

[15] Drndić M.; Jarosz M. V.; Morgan N. Y.; Kastner M. A.; Bawendi M. G. *J. Appl. Phys.* 2002, 92, 7498- 7503.

[16] Xin S. H.; Wang P.D.; Yin A.; Kim C.; Dobrowolska M.; Merz J. L.; Furdyna J. K. *Appl. Phys. Lett.* 1996, 69, 3884- 3886.

[17] Avramescu A.; Ueta A.; Uesugi K.; Suemune I. *J. Crystal Growth* 2000, 214/215, 125- 129.

[18] Kim S. W.; Kotani T.; Ueda M.; Fujita Shizuo; Fujita Shigeo *Appl. Phys. Lett.* 2003, 83, 3593- 3595.

[19] Eaglesham D. J.; Cerullo M. *Phys. Rev. Lett.* 1990, 64, 1943- 1946.

[20] Daruka I.; Barabási A. L. *Phys. Rev. Lett.* 1997, 79, 3708- 3711.

[21] Tawara T.; Tanaka S.; Kumano H.; Suemune I. *Appl. Phys. Lett.* 1999, 75, 235- 237.

[22] Kim Y. G.; Joh Y. S.; Song J. H.; Sim E. D.; Baek K. S.; Chang S. K.; Lee J. I, *Appl. Phys. Lett.* 2004, 85, 2056- 2058.

[23] Mackowski S. *Thin Solid Films*, 2002, 412, 96- 106.

[24] Mackowski S.; Karczewski G.; Kossut J.; Sek G.; Misiewicz J.; Prechtl G.; Heiss W. *PHYSICA E* 2002, 12, 503- 506.
[25] Bradford C.; Urbaszek B.; Funato M.; Graham T. C. M.; McGhee E. J.; Warburton R. J.; Prior K. A.; Cavenett B. *J. Crystal Growth* 2003, 251, 581- 585.
[26] Kim C. S.; Kim M.; Furdyna J. K.; Dobrowolska M.; Lee S.; Rho H.; Smith L. M.; Jackson H. E.; James E. M.; Xin Y.; Browning N. D. *Phys. Rev. Lett.* 2000, 85, 1124- 1127.
[27] Shan C. X.; Fan X. W.; Zhang J. Y.; Zhang Z. Z.; Wang X. H.; Lu Y. M.; Liu Y. C.; Shen D. Z.; Lu S. Z. *J. Crystal Growth* 2004, 265, 541- 547.
[28] Yang Y.; Shen D. Z.; Zhang J. Y.; Fan X. W.; Zhen Z. H.; Zhao X. W.; Zhao D. X.; Liu Y. N. *J. Crystal Growth* 2000, 220, 286- 290.
[29] Shan C. X.; Fan X. W.; Zhang J. Y.; Zhang Z. Z.; Li B. S.; Lu Y. M.; Liu Y. C.; Shen D. Z.; Kong X. G. *J. Vac. Sci. Technol. B*, 2002, 20, 1102- 1106.
[30] Kurtz E.; Schmidt M.; Litvinov D.; Dal Don B.; Dianoux R.; Zhao Hui; Kalt H.; Roseñauer A.; Gerthsen D.; Klingshirn C. *Phys. Stat. Sol.* (b) 2001, 229, 519- 522.
[31] Hinzer K.; Bayer M.; McCaffrey J. P.; Hawrylak P.; Korkusinski M.; Stern O.; Wasilewski Z. R.; Fafard S.; Forchel A. *Phys. Stat. Sol.* (b) 2001, 224, 385- 392.
[32] Seufert J.; Rambach M.; Bacher G.; Forchel A.; Keim M.; Ivanov S.; Waag A.; Landwehr G. *Phys. Rev. B* 2000, 62, 12609- 12612.
[33] Keim M.; Korn M.; Seufert J.; Bacher G.; Forchel A.; Landwehr G.; Ivanov S.; Sorokin S.; Sitnikova A. A.; Shubina T. V.; Toropov A.; Waag A. *J. Appl. Phys.* 2000, 88, 7051- 7055.
[34] Ivanov S. V.; Toropov A. A.; Shubina T. V.; Sorokin S. V.; Kyutt R. N.; Sitnikova A. A.; Solnyshkov D. D.; Nekrutkina O. V. *Phys. Stat. Sol.* (b) 2004, 241, 531–537.
[35] Ivanov S. V. *J. Alloy. Compound.* 2004, 371, 15- 19.
[36] Shubina T. V.; Ivanov S. V.; Toropov A. A.; Sorokin S. V.; Lebedev A. V.; Kyutt R. N.; Solnyshkov D. D.; Pozina G. R.; Bergman J. P.; Monemar B.; Willander M.; Waag A.; Landwehr G. *Phys. Stat. Sol.* (b) 2002, 229, 489- 492.
[37] Zhang B. P.; Yasuda T.; Segawa Y.; Yaguchi H.; Onabe K.; Edamatsu E.; Itoh T. *Appl. Phys. Lett.* 1997, 70, 2413- 2415.
[38] Zhang B. P.; Li Y. Q.; Yasuda T.; Segawa Y.; Edamatsu K.; Itoh T. *J. Crystal Growth* 2000, 214, 765- 769.
[39] Zhang B. P.; Li Y. Q.; Yasuda T.; Wang W. X.; Segawa Y.; Edamatsu K.; Itoh T. *Appl. Phys. Lett.* 1998, 73, 1266- 1268.
[40] Yasuda T.; Zhang B. P.; Segawa Y. *Electronics and Communications in Japan, Part II-Electronics* 1999, 82, 51- 60.
[41] Liao M. C. H.; Chang Y. H.; Chen Y. F.; Hsu J. W.; Lin J. M.; Chou W.C. *Appl. Phys. Lett.* 1997, 70, 2256- 2258.
[42] Liao M. C. H.; Chang Y. H.; Tsai C. C.; Chieng M. H.; Chen Y. F. *J. Appl. Phys.* 1999, 86, 4694- 4696.
[43] Chang Y. H.; Chieng M. H.; Tsai C. C.; Liao M. C. H.; Chen Y. F. *J. Electron. Mater.* 2000, 29, 173- 176.
[44] Seifert W.; Carlsson N.; Johansson J.; Pistol M. E.; Samuelson L. *J. Crystal Growth* 1997, 170, 39- 46.
[45] Yang Y.; Shen D. Z.; Zhang J. Y.; Fan X. W.; Zhen Z. H.; Zhao X. W.; Zhao D. X.; Liu Y. N. *J. Crystal Growth.* 2001, 225, 431- 434.

[46] Yang Y.; Shen D. Z.; Zhang J. Y.; Fan X. W.; Li B. S.; Lu Y. M.; Liu Y. C.; Liu Y. N. *J. Crystal Growth* 2001, 233, 785 – 790.
[47] Kim J. C.; Rho H.; Smith L. M.; Jackson E.; Lee S. Dobrowolska M.; Furdyna J. K. *Appl. Phys. Lett.* 1999, 75, 214- 216.
[48] Sebald K.; Michler P.; Passow T.; Hommel D.; Bacher G.; Forchel A. *Appl. Phys. Lett.* 2002, 81, 2920- 2922.
[49] Ulrich S. M.; Strauf S.; Michler P.; Bacher G.; Forchel A. *Appl. Phys. Lett.* 2003, 83, 1848- 1850.
[50] Fonoberov V. A.; Balandin A. A. *Appl. Phys. Lett.* 2004, 85, 5971- 5973.
[51] Mukai K.; Nakata Y.; Otsubo K.; Sugawara M.; Yokoyama N.; Ishikawa H. *Appl. Phys. Lett.* 2000, 76, 3349- 3351.
[52] Kobayashi M.; Nakamura S.; Wakao K.; Yoshikawa A. *J. Vac. Sci. Technol. B* 1997, 16, 1316- 1320.
[53] Kobayashi M.; Nakamura S.; Kitamura K.; Umeya H.; Jia A.; Yoshikawa A.; Shimotomai M.; Kato Y.; Takshashi K. *J. Vac. Sci. Technol. B* 1999, 17, 2005- 2008.
[54] Koayashi M.; Kitamura K.; Umeya H.; Jia A. W.; Yoshikawa A.; Shimotomai M.; Kato Y.; Takahashi K. *J. Vac. Sci. Technol. B* 2000, 18, 1684- 1687.
[55] Kitamura K.; Umeya H.; Jia A.; Shimotomai M.; Kato Y.; Kobayashi M.; Yoshikawa A.; Takahashi K. *J. Crystal Growth* 2000, 214/215, 680- 683.
[56] Nakamura S.; Kitamura K.; Umeya H.; Jia A.; Kobayashi M.; Yoshikawa A.; Shimotomai M.; Kato Y.; Takahashi K. *Electron. Lett.* 1998, 34, 2435- 2436.
[57] Passow T.; Klude M.; Leonardi K.; Hommel D. *Proc.25th Internat. Conf. Phys. Semicond.* (ICPS), Osaka (Japan), 2000.
[58] Matsumura N.; Endo H.; Saraie J. *Phys. Stat. Sol.* (b) 2002, 229, 1039- 1042.
[59] Ivanov S. V.; Toropov A. A.; Sorokin S. V.; Shubina T. V.; Sedova I. V.; Sitnikova A. A.; Kop'ev P. S.; Alferov Z. I.; Lugauer H. J.; Reuscher G.; Keim M.; Fischer F.; Waag A.; Landwehr G. *Appl. Phys. Lett.* 1999, 74, 498- 500.
[60] Klude M.; Passow T.; Heinke H.; Hommel D. *Phys. Stat. Sol.* (b) 2002, 229,1029-1032.
[61] Suemune I.; Yoshida K.; Kumano H.; Tawara T.; Ueta A.; Tanaka S. *J. Crystal Growth* 2003, 248, 301- 309.
[62] Xia J. B. *Physics* 1998, 27, 141- 145 (in Chinese).
[63] Kratzert P. R.; Rabe M.; Henneberger F. *Phys. Rev. Lett.* 1999, 83, 239- 240.
[64] Lee S.; Daruka L.; Kim C. S.; Barabási A. L.; Merz J. L.; Furdyna J. K. *Phys. Rev. Lett.* 1998, 81, 3479- 3482.
[65] Roest A. L.; Kelly J. J.; Vanmaekelbergh D.; Meulenkamp E. A. *Phys. Rev. Lett.* 2002, 89, 036801 1- 4.
[66] Roest A. L.; Kelly J. J.; Vanmaekelbergh D. *Appl. Phys. Lett.* 2003, 83, 5530-5532.
[67] Coe S.; Woo W. K.; Bawendi M.; Bulovic V. *Nature* 2002, 420, 800- 803.
[68] Tsutsui T. *Nature* 2002, 420, 752- 754.

INDEX

A

B

C

D

E

F

G

N

O

P

Q

U

V

W

X

Y

Z